# THE OPTICAL MICROSCOPE MANUAL

# THE OPTICAL MICROSCOPE MANUAL

Past and Present Uses and Techniques

BRIAN J. FORD

DAVID & CHARLES : NEWTON ABBOT
CRANE, RUSSAK : NEW YORK

© BRIAN J. FORD 1973

First published 1973 by
David & Charles (Holdings) Limited
South Devon House   Newton Abbot   Devon

(ISBN 0-7153-5862-6)

Published in the United States of America by
Crane, Russak & Company, Inc.
52 Vanderbilt Avenue
New York, New York 10017

(ISBN 0-8448-0157-7)

Library of Congress Catalogue Card Number 72-94378

All rights reserved. No part of this
publication may be reproduced, stored
in a retrieval system, or transmitted,
in any form or by any means, electronic,
mechanical, photocopying, recording or
otherwise, without the prior permission
of David & Charles (Holdings) Limited

Set in 11/13pt Plantin
and printed in Great Britain
by W J Holman Limited Dawlish

# CONTENTS

|   |   | page |
|---|---|---|
| 1 | The Pioneers | 9 |
| 2 | Mass-production Microscopy | 26 |
| 3 | The Era of Brass and Glass | 42 |
| 4 | The 'Modern Microscope' arrives | 77 |

The Research Microscope—'Miniaturisation' in Design—The Largest Optical Microscopes—Phase Contrast—Interference Microscope—Lasers and Holomicrography—Dark-ground Microscopy—The Invisible Wavelengths—Fluorescence Microscopy—The State of the Art

| 5 | The Microscope in Action | 98 |
|---|---|---|

Setting up the Instrument—Types of Illuminant—The Mirror—Condenser Systems—Slide—Immersion Oil—Objectives—The Microscope Body—Eyepieces—The Stand—The Zoom Attachment

| 6 | The Microscope in Use | 137 |
|---|---|---|

Student Microscopes—Binocular Research—Microscope—Vibration—Adjusting the Illumination—The Condenser-Lamp Distance

| 7 | The Specimen | 156 |
|---|---|---|

Examining the Specimen—The Solid Specimen—The Particulate Specimen—Biological Material-Living—Biological Material-Dead (Temporary Mounts)—Biological Material-Dead (Permanent Preparations)

| Bibliography | 197 |
|---|---|
| Index | 199 |

# LIST OF ILLUSTRATIONS

| TEXT FIGURES | page |
|---|---|
| 1a  Diagram of light beam normal to glass block | 10 |
| 1b  Diagram of light beam passing at an angle into glass, showing refraction | 10 |
| 2a  Diagram of light rays 'spread' by a biconcave lens | 14 |
| 2b  Idealised diagram of light focussed by a biconvex lens | 14 |
| 2c  Diagram showing relationship of image and object size | 15 |
| 3   Drawing of the typical form of microscope constructed by Leeuwenhoek | 27 |
| 4   An early microscope of Culpepper-Scarlett form dated circa 1730 | 38 |
| 5   Diagram of chromatic aberration | 50 |
| 6   Passage of rays through compound microscope | 78 |
| 7   Idealised diagram of a split-beam interference microscope | 89 |
| 8a  High-speed scanning of an epithelial cell | 91 |
| 8b  Computer print-out of percentage transmission in different regions of the epithelial cell | 92 |
| 9   Evolution of the microscope preparation | 158 |
| 10  Transverse section of a typical slide in use | 192 |

PLATES

| | |
|---|---|
| A rare engraving of Leeuwenhoek's microscopical methods, showing the typical microscope | 33 |
| Drawing by Hooke (from *Micrographia*) showing his method of working | 34 |

| | |
|---|---|
| Plate from Hooke's *Micrographia* showing a magnified drawing of woven cloth | 34 |
| A very fine engraving by Tinney of Cuff's 'Double Microscope' dating from 1744 | 51 |
| Contemporary engraving of the Wilson screw-barrel ready for use (circa 1720) | 52 |
| A picture of the Powell & Lealand No 1 stand in use | 52 |
| Four plant stem sections (one a petiole) photographed by Fox Talbot (circa 1841) | 85 |
| A popular Victorian pastime—the mounting of siliceous spicules of sponges in intricate patterns | 85 |
| Another example of spicule mounting | 85 |
| Microscope slides from 1720-1920 | 86 |
| Early eighteenth- and early twentieth-century slides | 86 |
| The Burch reflector in use at the Chester Beatty Institute, London | 103 |
| A research microscope in action—the Vickers Photoplan equipped for transmission photomicrography | 104 |
| Specimen assessment automated by the use of a scanning microscope | 104 |

CHAPTER ONE

# THE PIONEERS

THE EARLIEST microscopists were, in today's terms, nothing of the sort. They did not have microscopes, but only simple lenses; and they did not employ any microtechnique in preparing their specimens for examination—they looked at them as they were. Most of the specimens examined were lifeless, a fragment of dried moss or a swatted fly; and probably the first form of technique used to prepare a specimen for examination was simply to impale it on a pin.

It was in 1614 that Giovanni du Pont—who had built telescopes—wrote for the first time of the observation that a microscope could be made by using a telescope back-to-front, and increasing the distance between the lenses. His telescopes, he wrote, were typically two feet long; but used backwards they could show him small objects that 'we could not distinguish because of their small size'. In this mode of use, the length between the two lenses was two arm's-lengths, he wrote, instead of two feet. This was much the same as the type of simple microscope (no more than a high-powered convex objective lens and a weak eyepiece) first suggested by Kepeer in 1611, and the observation of 'small objects' too tiny to be seen by the naked eye clearly does not call for the use of any microtechnique.

In 1637 Descartes first showed a diagram of a microscope with a pointed stylus to support the object. His idea was for the use of a concave mirror to reflect the sun's rays up and on to the

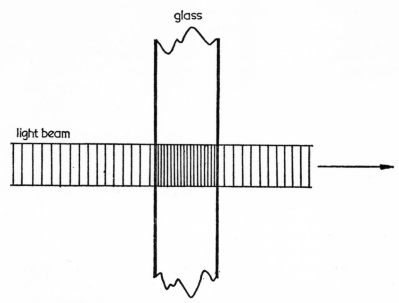

Fig 1a  Diagram of light beam normal to glass block

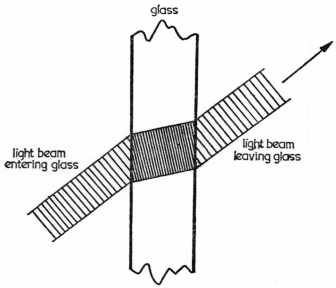

Fig 1b  Diagram of light beam passing at an angle into glass, showing refraction

object, impaled on the point of the stylus; and in this way the first actual techniques of microscopy were recorded. Descartes's diagrams introduced another important idea into pioneer microscopy—the use of transmitted, rather than reflected, light. The earliest observations were carried out on objects lit from above, and merely standing on an opaque background. But Descartes showed at least one diagram in which a condenser lens was included—and this was meant for concentrating the light on to a specimen in much the same way as in conventional microscopy today. It is unlikely that Descartes ever built any of these instruments, however. Most students of the period believe that it would have been beyond the technical resources available to him at that time.

The microscope remained a mere curiosity, no more than a toy in essence, until the 1660s. It was only then that microscopy as a recognisable discipline began to appear. That event was largely due to the inquisitiveness and ingenuity of one man—Robert Hooke. He was born in 1635 in Freshford, Isle of Wight, England, and had an excellent early training as apprentice to a London portrait artist. He threw this career up after a little while, but the craft of accurate and painstaking illustration was to serve him well when he embarked on a career as an investigator of the microscopy of familiar objects.

By 1660 he had clearly become interested in the minute details of conventional structures that could be revealed by the use of lenses, and with his appointment as Curator of Experiments at the newly-formed Royal Society, this interest found a new impetus. So intrigued were the members by his inquisitiveness and by the revelations of minute structure with which he could present them, that on 1 April 1663—All Fools' Day—he was instructed to present one new microscopical demonstration each week for the delectation of the members.

His first demonstration, a week later, was a piece of conventional magnification. He showed a specimen of moss, drawn with infinite care, as seen impaled on a pin-point in the customary manner. For the most part, like the majority of his demonstrations, the interest of his results lay in the mere size of the image. No one was greatly interested in any new forms of life that were invisible to the naked eye, only in the sight of a conventional, familiar object made to look larger—an important distinction from the later microscopists. But there was an important result from this initial observation. Five of the leaflets he drew in great detail. They showed not just the shading lines and the (incorrect) indication of veins—a convention wrongly taken from the leaves of higher plants—but a regular pattern of small, rectangular markings like bricks in a wall. Hooke was observing the cellular structure of living organisms for the first time. During the following week he took the observations a stage further.

He turned his attention to another plant tissue—cork. He found that direct observation did not show him much, and so he took a prophetic step: he carefully shaved off a thin section with a sharp razor edge. This was the first recorded cutting of a section for microscopical examination—and sections of materials, whether plant or animal, mineral or artificial fibre, are now a mainstay of microscopical technique. Unlike today's preparations, however, which are mounted in a transparent medium for examination by transmitted light, Hooke viewed his cork sections against a black background, with reflected illumination. But even if the concept of the microscopical section had a long way to go, assuredly this was the first step in that direction.

The unfolding of the concept of a section in Hooke's mind is clearly evident in his description of the way in which he cut this, the first ever to be described. In his book *Micrographia*, published two years later, he recounts how he took the blade of a pen-knife, and sharpened it very carefully until its edge was as good as that of any razor. With this he removed a piece from a bottle cork, and placed it under his microscope. Looking very carefully he could see the regular pattern of minute pores (no doubt similar, in his mind, to the appearance of the moss leaflets he had seen a week before). But his curiosity was aroused, he wrote; and he realised that there was every chance that the entire structure of the cork was porous. After all, he reasoned, that would explain why it is so light a substance, so water-tight, and so buoyant. It was this that encouraged him to examine the pores in greater detail and so, from the smooth, flat surface from which he had already removed his first slice of the cork, he cut off an 'exceeding thin piece of it' and placed it on a black object plate. The light he cast on the specimen from a large plano-convex lens (probably cast, and not ground). It was then that he could perceive the cells more plainly. This was the birth—albeit a premature event in many ways—of histology, the microscopic study of tissues.

In Hooke's day the most important activity for the pioneer microscopist was not (as it is now) the preparation of permanent mounts for examination, the staining of tissues and the sectioning of specimens—indeed most of our modern techniques were undreamed of at that time. Instead, it was centred on the time-consuming and tedious task of making apparatus, assembling equipment and grinding lens elements. Hooke used to commission his microscopes from craftsmen who could work to a design, even though they probably knew not the first thing about microscopes, science or anything else apart from artisanship. But even this left a great deal of work for his hands alone. This paraphrased account dating from 1663 gives us some insight into the microscopy of the time:

> The microscope I most generally made use of had a tube for the most part not above six or seven inches long, though because it contained four drawtubes it could naturally be greatly lengthened if necessary. The tubes contained three lenses: a small objective, a thin eye lens and a stronger eyepiece at the top. However, I only used this to examine whole specimens, for when it was necessary to examine the minute details of an object more closely, I would take out the middle lens and use only the two remaining. Always, the fewer refractions there are, the brighter the image appears to be. Certainly, if we could make a microscope with only one lens, it follows that the image would be the best of all.
> And so it is best to take a piece of broken Venetian glass of

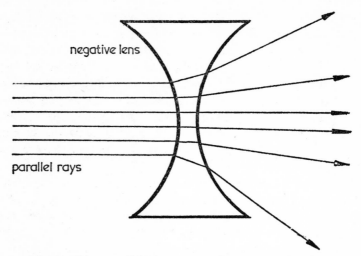

Fig 2a  Diagram of light rays 'spread' by a biconcave lens

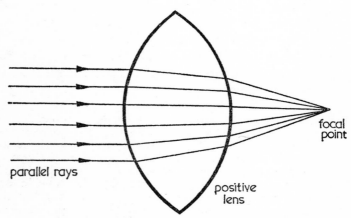

Fig 2b  Idealised diagram of light focussed by a biconvex lens

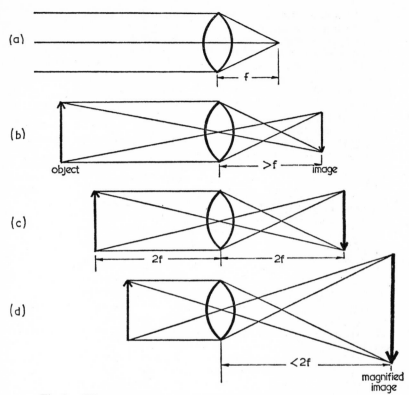

Fig 2c Diagram showing relationship of image and object size

high quality and draw it out after heating in the flame to form fine threads. When one of these is heated in the flame, the glass melts into a bead which can be carefully controlled until it is of manageable proportions, hanging on the end of the thread. Several of these can be stuck with sealing wax on to a stick, so that the pointed threads stick upwards. Rubbing against a whetstone will remove and flatten the surface somewhat, and then the polishing can proceed using a little tripoli paste (a mixture of the mineral tripolite) on a smooth metal plate.

The lenses thus formed can be used as magnifiers on their own. Yet because of the inconvenience of this and their smallness, and yet wishing to have no more than two refractions (ie lenses), I obtained a brass tube in the small end of which I fitted with wax a good plano-convex lens with the convex side towards the object, and into the wider end I fixed, also with wax, a large plano-convex glass with the convex side towards my eye. Through a small hole in the side of the tube I poured water to entirely fill the barrel of the microscope, and closed the aperture with a screw. I found the image appeared to be brighter than when the intermediate space was filled with air, but because of the inconvenience of the idea I have made little use of it.

My way of mounting the normal microscope and object on to a pedestal was to fit a small pillar with an iron arm that could slide up or down. It could be fixed in any desired position by means of a small screw. On the arm was a ball and socket joint into which the microscope tube itself was fixed. By the use of this apparatus I could fix the microscope in whatever position I desired and adjust it to suit any kind of object. For placing the object I fixed a small brass staple with a small pillar, three-quarters of an inch high, on the side of it, through the top of which was a small iron pin. I could move this into any position without disturbing the position of the microscope tube. This was the microscope I most frequently made use of.

I have made several other trials with other kinds of microscopes which were quite different from ordinary spherical lenses. I have made a microscope with one piece of glass, both of whose surfaces were planes; I have made another with a plano-concave lens only [It is difficult to see what he means by these descriptions—both ideas seem to be quite impracticable] and several others that are reflecting microscopes.

I have made lenses with water, transparent gums, resins, salts, compounds of arsenic and various oils and with several other liquids. And indeed the whole subject gives endless possibilities. However, I find that the best design is based on the system of two lenses outlined above.

This account shows how much the microscopist of the time was geared to self-reliance. And it illustrates the degree of innate craftsmanship that Hooke could use in solving a problem. The grinding of a lens as he describes it necessitates the orientation of the lens in exactly the same position on the polishing surface throughout the grinding process—which is a very difficult task to undertake. But by the simple expedient of mounting a number of globules together on a plate of sealing-wax and applying them to the grinding surface, he hit on a splendidly foolproof technique. Obviously, once the two surfaces are laid together they will face each other like a studded boot standing on a piece of glass. The orientation of the lens globules (the 'studs') will remain the same throughout the grinding operation and errors of orientation on the part of the operator are thereby removed from the variable factors that might spoil the quality of the results.

The manufacture of the microscope body was, of course, a matter for the craftsman. Christopher Cock, who at that time had premises in Long Acre (now near London's Covent Garden tube station) was particularly recommended by Hooke. Dr Henry Power—a contemporary—used an instrument made for him by a craftsman named Reeves, who also made the instrument that the diarist Samuel Pepys purchased whilst he was at the Navy Office. The price of such an instrument was between £5.50 and £8.50—not by any means a low cost for the time.

The amateur microscopist, once he had an instrument, would spend much time in manipulating specimens on slips of mica or on glass fragments—perhaps even in making his own objective lenses in the manner described by Hooke. The beginner, no doubt, would use the ready-made lenses that he could purchase with the microscope and might content himself by impaling insects on the pointed iron pin until the novelty wore off, just as the purchasers of 'student' toy microscopes do today. But microscopy as a subject embraced a broad range of techniques. It was a very trial-and-error business.

Hooke, in using dry specimens, was soon to fall foul of these primitive techniques. Towards the end of April 1663, he turned his attention to a pin-mould (*Mucor*, perhaps) which he found on a decayed leather book-binding. These fungi form aerial spore-bearing hypae from a tangled mass of mycelium. At the apex of each of these branches forms a rounded swelling which matures into a spherical sporangium. Inside it, in large numbers, spores are produced. The breakdown of the sporangium releases them (occasionally, as I have often observed in species of the related genus *Rhizobium*, the spores are ejected by an explosive collapse of the sporangium wall) and they are always plainly visible in mature specimens of the fungus.

But Hooke's use of dried specimens let him down. The fungus bodies were too old; those that had dehisced, or dried, had lost their spore complement entirely, whilst those that had not matured remained dead and dry before his lens. The result was that he observed no spores at all. To the modern microscopist, who uses fresh material mounted in a fluid medium—which makes spore loss impossible—it is inconceivable that a mature colony of this sort would not be liberally covered with newly-released spores. But Hooke wrote: 'Some [of the sporangial structures] were small and short, as seeming to have been but newly sprung up, of these the balls were for the most part round, others were bigger, and taller, as being perhaps of a longer growth, and of these, for the most part, the heads were broken and some much wasted; what these heads contained I could not perceive'.

He concluded that the moulds required no 'seminal property' —spores, in other words—to reproduce, and speculated on the possibility that mushrooms and their relatives arose chemically much as the feathery crystals seen in saturated solutions or forming as frost on a window-pane. Had he been just a little more diligent, and turned his attention to a mature—but living— specimen of mould, he would doubtless have observed spore formation. The analogy with seeds would have been obvious to him, and he might then have laid the foundations for those who were later to preach against the doctrine of spontaneous generation. If these minuscule plant forms were reproduced by tiny seeds, then surely the other forms of life that others were to discover later must have a similar phase in their life-cycle, or something similar . . . but this idea was not to dawn.

Yet Hooke was aware of the benefits of using specimens mounted in fluid. For the study of muscles and tendons in small insects he recommended the mounting of the specimen 'in a liquor, such as water or a very clear oyl.' This is an essential principle of today's routine microscopy—it is unfortunate that Hooke did not extend the principle to other types of specimen, as it would have greatly facilitated many of his observations.

We may gather something of the way in which he worked at that time by referring to his notes (later published in *Micrographia*). He would set up a firm table in a room with one window only, facing the south; the edge of the table would be three or four feet from the window itself. On this he stood his microscope. There were several alternatives for the technique of illumination of the specimen. His usual method, it seems, was to set up a round flask of water (to act as a lens) or a very thick planoconvex lens of glass, mounted on a wooden clamp in such a position as to concentrate the daylight in a pool around the specimen stage. This was generally enough, he wrote, to light the specimen so brightly as to enable the microscopist to obtain twice the magnification possible with no more than direct daylight. If the day was sunny, Hooke would use one of two alternative methods for diffusing the sunlight.

In one technique he would set up the condenser apparatus as before, but between the lens and the sunlight he would insert a sheet of oiled or waxed paper. The effect of this translucent filter was to give a more diffuse pool of light than the sunlight itself would provide. Not only did this remove the glare 'which the immediate light of the sun is very apt to create in most objects', but it prevented over-heating of the specimen. The direct sunlight that could be concentrated through such a powerful condenser lens as Hooke describes would be enough to set fire to most specimens in a matter of minutes.

A second, more sophisticated technique was to reflect the light from a mirror surface. Hooke would take a small piece of a broken looking-glass and give it a matt surface by rubbing it for a little while on a flat tool surface sprinkled with fine sand. This, set up in such a way as to reflect the sun's rays through the condenser lenses as before, acted as a diffuser, too, and gave more reliable service than the sheet of translucent paper. He also found that the mirror reflected more light than the oiled paper would transmit, which was an added advantage.

None the less, Robert Hooke soon found that the weather in the springtime of 1663 was too changeable to rely on. So he decided to build himself an artificial light-source that did away with his dependence on sunlight, and which enabled him to work in the evenings as well. He found an alchemist's stand with two projecting arms that could be fixed in any position by means of a pair of screws, clearly shown in his drawing of the apparatus (see plate, p34). On the uppermost bracket he mounted an oil lamp. The light from the flame was concentrated by a round flask filled with brine. (Hooke found that his measurements showed salt water to be more refractile than fresh, and perhaps this is why he filled the condenser with brine. It is equally likely that he did so in order to prevent the growth of algae, which would quickly happen in such a container exposed to daylight, and which would necessitate frequent refilling and cleaning of the flask.)

To eliminate the tedious and delicate manoeuvre involved in re-setting the position of the lamp and the large flask, in order to cast the greatest concentration of light on the specimen, Hooke arranged for a jointed arm to be fixed to the lower bracket, on the end of which a solid glass lens was mounted (labelled I in the diagram). By moving this at will he could easily make small changes in the direction of the lighting and could control the intensity of the illumination.

He was also able to effect some further control of lighting by moving the specimen itself. Thus, when he examined a sharpened razor edge he tilted the blade until it reflected light from its cutting edge, causing it to show as a bright line across the field of view—an easily visible line of demarcation revealing the edge of the metal.

He therefore had several different means of altering the illumination of a given field of view and, more than this, he was able to match the lighting conditions to the nature of the specimen or (in the case of the razor) to emphasise a particular aspect of the structure. This flexibility in principle underlies a deal of modern microtechnique.

Hooke's detailed account of his observations of the head of a fly reveal his dependence on the use of different lighting conditions to reveal details of structure. He described how he removed the head of a fly and impaled it on the point of the needle of his microscope. 'Then I examined it according to my usual manner', he went on, 'by varying the degrees of light and altering its position to each kinde of light...' Clearly he realised how a slight change in the illumination of a solid object could alter the observer's visual interpretation—an important concept.

When examining opaque structures by reflected light, Hooke placed them either on the point of the needle of his instrument or on a specimen plate. In either event he chose the hue of the plate so that it would provide a suitable contrast for the specimen—thus a white plate would be appropriate for an insect; whilst the pale tint of his thin cork sections was best revealed by a blackened plate. Once or twice, when examining snowflakes caught as they fell, he used a piece of black cloth or an old 'black Hatt'—the fibrous nature of this substrate would keep the snowflake out of contact with the bulk of the specimen holder, and the low conductivity would prevent rapid melting. In this way he obtained some startlingly clear views of snowflakes, and a similar technique is favoured today.

And what of his technique for examining fluid specimens? Many writers have referred to his use of 'glass' plates pressed together (in the manner of a modern-day slide and coverslip); but my belief is that he principally used mica flakes. He referred to them as 'Muscovy glass', and white mica is still known to this day as muscovite. He would take a specimen of this material and separate flakes of it by pressing lightly against the edge of the grain with a needle. In this way, he wrote, it was possible to obtain very fine flakes of high transparency, and he selected pieces for use that were about an inch across. These, placed over a hole some $\frac{3}{4}$ inch in diameter in the middle of the microscope stage, were an ideal support for the examination of 'liquors'.

On several occasions Hooke tried to secure a living insect on just such an object plate, by coating the plate with 'Wax or Glew' so that the insect's feet were firmly anchored. That did not work, he writes, for the insect 'would twist and wind its body, that I could not any wayes get a good view of it'. He soon resorted to the use of alcohol as a means of killing the specimen by immersion—but he did not proceed to examine it in its fluid mountant. Rather he would leave it lying on a piece of paper until the alcohol had evaporated, before proceeding to mount it on the pin point in the usual way.

Yet he did discover the benefits of using an immersion objective lens—ie a system in which the preparation coverslip and the objective lens of the microscope are connected by means of a drop of fluid between them. He once spoke of using a 'liquor' with a refractive index similar to that of glass as a means of bridging the gap, and rightly drew attention to many of the possible advantages of this system. Although he recognised this to be a valuable aid to the microscopist and an important technique in microscopy, he seems to have utilised it infrequently.

Another technique that he did not exploit, though it was available, was that of tissue injection. This was a means of distinguishing the intricate reticulum of blood vessels in the internal organs. Coloured liquids were introduced into arteries through a hollow needle. As the fluid spread along the narrowing dividing blood channels it gave them a distinctive appearance that readily distinguished them from the surrounding tissues. It was a method developed by Sir Christopher Wren, whom we like to think of as an architect, immortalised in the splendour of St Paul's Cathedral, but who was also a gifted physician. Hooke described the technique as 'the ingenious Invention of that Excellent person, Doctor *Wren*'.

In truth the technique was being used elsewhere at that time. In Bologna, Italy, a young doctor in his early 30s had been studying animal physiology. It seems that he used fresh material on slips of mica for his observations, which centered on the minute structure of frog lung. This man was Marcello Malpighi, and he quickly realised the benefits of blood vessel injection as a means of demonstrating their course. Many of his studies were based on an examination of living or fresh lung tissue, a process which he found to be difficult and tedious. However, this is noteworthy in itself as it is the first known application of microscopy to living anatomy.

Malpighi discovered the capillary circulation of the lung alveoli and believed this to be an important piece of research. It was, for it provided one of the missing links in the theories of blood circulation. He went further than the examination of fresh lung tissue and the injection of the blood vessels, for he even managed to make some permanent preparations of the alveoli. To do this he simply allowed fragments of lung to dry, and he then found that the fine brownish pattern of the dried blood vessels could be easily seen. In all probability he made his first observation of dried lung by accident, having left a wet preparation in place overnight only to find it hard and dry next day. Be that as it may it remains the earliest example of a permanent histological preparation, albeit of a very crude nature.

Later he moved on to discover, by careful examination of dissected specimens, the pattern of circulation in the kidney. To this day the 'filtration' elements in the kidney are known as Malpighian corpuscles. And the *stratum spinosum* of the skin is still sometimes referred to as the Malpighian layer, along with characteristic nodules in the spleen which are named Malpighian corpuscles too. Clearly the abilities of this little-known Italian microscopist deserve to be better known, and it is worthy of note that all this work was carried out by the viewing of cut surfaces of the organs concerned (apart from the lung, which is essentially membranous in structure), with the aid of primitive injection techniques to help outline the blood vessels. Without any stains or other reagents, and knowing nothing about the potentialities of thin sectioning as an aid to the revelation of fine structure, he carried out some surprisingly detailed work. Happily his powers of observation and his manual dexterity made up for the crude microscopical techniques that were all he knew.

At about the same time, in Holland, a naturalist named J. Swammerdam was using an injection technique to aid his dissections of insects (particularly the bee *Apis mellifera*). He used mercury as well, but found molten candle wax to be a useful aid: it would readily penetrate the structures he was trying to inspect and then, having distended them and made them far easier to see, it would solidify. It is quite possible that this technique could still find applications today, though it is never used.

Swammerdam also added a little to the growing number of techniques of microscopy, by dissecting under water. This is still a favourite method for the microdissection of small specimens, as it allows the organs to float and the partial buoyancy encourages them to separate, thus aiding the clarity of the dissection. Insects such as these are small and microdissection of their internal organs calls for care and precision. Swammerdam was a master at it.

Though Hooke—because of the legendary standards of excellence in his drawings—is rightly regarded as the first documenter of the microscopic world, it should be borne in mind that he mainly specialised in enlarging details that were already familiar (the louse, the flea, the sting of a bee etc). Swammerdam's dissections were breaking new ground in the field of microscopy in the period 1663 onwards, roughly the same period as Hooke was active, and these detailed anatomical studies of insects were certainly revolutionary. With the exception of the cork section described above and the observations on the tongue of a snail, Hooke performed no alterations in his specimens (such as dissection) before examining and drawing them, and so his contribution to the techniques of microscopy must be regarded as of lesser importance.

Whilst Swammerdam was busy with small zoological specimens, it was another British scientist, Nehemiah Grew, who was actively cutting plants to pieces under the lens in order to resolve the details of plant structure. Between 1670 and 1680 he amassed a formidable amount of detailed information which was published in 1682 as the *Anatomy of Plants*. So the breadth of technique at that time can be seen to cover a broad field of investigation. Hooke concentrated on magnification of structures and popularising his work; Swammerdam studied invertebrate anatomy as Malpighi unravelled many details of vertebrate histology; and Grew worked with higher plants.

Between them the principles of fluid mounting, sectioning, vascular injection of a contrast medium and the mounting of permanent preparations had all arisen. And microtechnique became established as a result.

CHAPTER TWO

# MASS-PRODUCTION MICROSCOPY

AT THE same time that microscopes were becoming popular in Britain, the first mass-producing microscopist was beginning to emerge. He was a Dutch draper, Anthony van Leeuwenhoek by name, and he lived and worked in Delft some three centuries ago.

Leeuwenhoek is something of a legend. It is widely taught that he was the father of microbiology, as indeed in many respects he was; that he pioneered high-power microscopy, which is undeniably true, and that he was a humble self-taught man who started virtually from scratch and discovered it all by trial and error... but did he?

Leeuwenhoek has been universally acclaimed as a master of pioneer microscopy, a tireless seeker after knowledge and a dedicated observer as well. His microscopes were made in a manner that was—in contradiction to the picture most generally accepted—very far from delicate and craftsmanlike. In the main they were no more than rough plates of metal only as large as a couple of postage stamps, rivetted together and perforated with a small hole. It was at this perforation, gripped between the two metal units, that the tiny single lens was mounted. A pin, often adjusted crudely by a home-made screw, carried the object at the focal distance from this lens and, by observing through the lens itself, squinting into the daylight, microscopic organisms could be seen. (See Fig 3.)

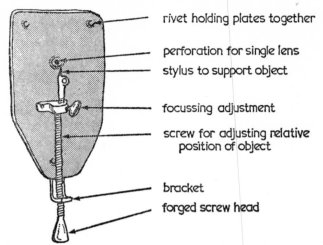

Fig 3  Drawing of the typical form of microscope constructed by Leeuwenhoek. The size of his instruments was generally small—smaller than they are usually depicted. Many of them are only a little larger than a typical rectangular postage stamp

This level of acquaintance is general amongst those who are interested in the history of biological microscopy. But it is not, I believe, quite as straightforward as it seems. It is very possible that Leeuwenhoek did not start from scratch at all. Before the legion supporters of his memory reach for their pens, perhaps I had better explain the basis for this belief.

Leeuwenhoek's biographer, Dobell, seems to reveal the key to the truth when he writes that Leeuwenhoek's first microscopic observations date from 1668. It was at that time that Hooke's *Micrographia* was at the height of its popularity, and Leeuwenhoek actually came to London *in that same year*. It seems a very fair assumption that—if he was nurturing an interest in microscopy at about that time—he would have seen a copy.

In Leeuwenhoek's account of the time, he wrote that he tried to 'penetrate' a specimen of chalk taken from cliffs at Gravesend and Rochester, in England. Dobell states that this 'evidently' means that he tried to study its microscopic structure. But the use of the phrase *te penetreren* could easily mean that he tried to understand its structure, that he tried to dig it with a point and saw the tiny flakes that came away, or that he attempted to examine it closely... the 'evident' allusion to microscopy that Dobell cites is without foundation. Much more likely is it that Leeuwenhoek was pondering over the subject in an idle moment, for he did not record the observations for six years and then did not emphasise any earlier interest in 'microscopy' at all.

There is an obvious reason why I believe he would have seen *Micrographia*. Leeuwenhoek was clearly interested in structure, or he would not have speculated idly on the nature of chalk. And he was a draper. As such the structure of cloth was clearly close to his interests—and Hooke's book, for the first time in a form available to a man in Leeuwenhoek's position, described the structure of cloth very fully. There are some excellent, detailed drawings of magnified cloth specimens in *Micrographia* and they are just what Leeuwenhoek would have found most interesting. (See plate, p34.) It follows that he might have read Hooke's description of the construction of microscopes—and this gives the most direct clue of all.

In the book, Hooke writes as follows:

> If one of these lenses be fixed with a little wax against a needle hole pricked through a thin plate of Brass, Lead, Pewter or any other metal, and an object placed very near be looked at through it, it will both magnify and make some objects more distinct than any of the large [ie compound] microscopes. But though they are exceedingly easily made, they are yet troublesome to use...

And that, though attention has not been drawn to this vital passage from the book before, is a *description of a Leeuwenhoek microscope*!

It seems very clear that his visit to London showed him something of the potentialities of microscopy, and the very book that contained the drawings of textiles under the microscope also included instructions for the manufacture of simple microscopes. There are many ways of mounting simple lenses, after all, and different microscopists used different methods. It would be the height of naïvety to conclude that Leeuwenhoek's adherence to Hook's specifications was due to some fortuitous coincidence.

But even though the manufacture of these little instruments was uncomplicated, it resulted in a magnifier that—in Hooke's estimation—could exceed the performance of the most expensive compound microscope then in existence. The impetus to Leeuwenhoek would have been considerable. With his talents as a handyman, and such a description of techniques already available (the production of lenses is described earlier, p15), it would have been obvious for him to commence his new hobby after he returned to Delft. And that is exactly what his own records suggest he did.

The sequence of events, we may postulate, is this:

1. Leeuwenhoek, at the age of 36, sails from Holland to London during a spell of relative peace between the Dutch and the British (the two nations were warring, on and off, during this period; but between 1667 and 1672 there was a truce—and Leeuwenhoek came to Britain in the first year of the peace, which suggests he was nurturing some particular desire behind the visit. What that might be is unknown).

2. He travels via Gravesend and Rochester and, as he is an inquisitive man, he is intrigued by the chalky cliffs in those places. Cliffs are themselves a novelty to a Dutchman used to the vast plains of his homeland, and the unnatural appearance of pure white chalk would make any such visitor in transit very likely to gather some samples and perhaps dig at them with a pocket knife. His later account suggests he did so.

3. Whilst in London he sees a copy of *Micrographia*, or has one drawn to his attention by a friend. Perhaps he stayed with a fellow-draper, a trader in textiles, or some such person, who might have known of the studies of cloth in Hooke's book. *This is the only event in the sequence of which there is not documentation.*

4. In *Micrographia* are drawings of fabrics, of obvious interest

to any inquisitive draper, instructions for grinding lenses; a design for a simple microscope; and an assurance that the results obtainable can exceed those of the most sophisticated compound instruments then available.

5. Some while after his return to Delft, Leeuwenhoek begins some experiments himself, which involve the manufacture of microscopes very similar indeed to those described by Hooke. The first record we have is in a letter about him written to the Royal Society by Renier de Graaf in April 1673, in which his recent interest in microscopy is mentioned. Leeuwenhoek had, said de Graaf, obtained some interesting results—and the observations included with the letter were Leeuwenhoek's studies of a mould, the mouthparts and sting of a bee, and *Pediculus* (the louse). *These are also specimens featured in Micrographia!*

He observed, in September 1678, a funguous infection of grasses. And in his description of the passage of sap through the plant's vessels he describes them, not as tubules, pipes, ducts or whatever; but as 'pores'—the same term used by Hooke in *Micrographia*.

Further coincidental examples are revealed in the list of specimens that he eventually bequeathed to the Royal Society. Many of them, including fleas, lice, and the eyes of insects, would have occurred to any amateur microscopist of that period and we must not draw undue conclusions from the fact that they are also in the pages of *Micrographia*. But some of them—such as the spinnerets of a spider and the fragments of metal struck as sparks with a flint—are very likely the result of direct inspiration. Certainly the list of low-power specimens is far too similar to the specimens in *Micrographia* to make mere 'coincidence' a likely explanation.

Leeuwenhoek was not the innovator we like to imagine. The widespread adulation of his memory does not make this an easy thesis to advance; but the evidence is plain enough. His methods and in many cases his specimens follow closely the dictates in Hooke's published work and, for all the undoubted craftsmanship with which Leeuwenhoek pressed ahead, there is not a scrap of evidence to suggest that he was a visionary, an inventor, or a philosophical genius. Indeed he has even been described quite recently as the 'discoverer of the microscope'—which is as sensible as calling Henry Ford the 'discoverer of transport'.

Certainly there are many who will decline to accept such an implied slur on the memory of Leeuwenhoek; but there is an undeniable mass of evidence—even if we have not seen it before—that points quite clearly in this different direction.

So, though it seems to me quite clear that we cannot accept Leeuwenhoek's historical place as a microscope innovator—and the term 'Leeuwenhoek microscope' might best be understood to refer to his version of a simple magnifier, and not to suggest that he was the only worker to adhere to this design—we must certainly acknowledge his place as a pioneer in microscopy. Leeuwenhoek was following Hooke's lead in many ways (even in his choice of specimen at first) but the work he later undertook founded many of our present-day microscopical techniques.

He prepared a great many sections of plant material, cut both as transverse and longitudinal sections (he did not distinguish between radial and tangential longitudinal orientations, however). Dobell, Leeuwenhoek's fanatically devoted biographer, suggests that he was one of the first, 'if not the very first', to study specimens sectioned with a razor. It is interesting to note that some of the specimens were sent to London and accepted by the Royal Society. The sections were in a little packet attached to one of his letters, and the specimen listed as No 1 is—predictably enough—'Cork'... another famous example from *Micrographia*. In this instance Hooke had pre-dated Leeuwenhoek by a good eleven years and, far from being a revolution in microscopy, by 1674 (the date of the letter that accompanied the specimens from Holland) it was virtually a standard specimen.

But his handling of liquid mounts was certainly original. In examining the colonial flagellate *Anthophysa vegetans*, for instance, he found it was best studied by allowing the water to drain out of the glass tube containing the specimen, allowing it to lie—stranded, as it were—against the glass. In this way, he recorded, it could most easily be seen by his draughtsman. (Leeuwenhoek was apparently a very poor artist; nearly all his drawings were done by others on his instruction.)

This is an intelligent approach to the problem, and quite significant in such early days of microscopy. I have used a similar technique to obtain permanent preparations of filamentous algae (including *Spirogyra, Zygnema* and *Cladophora*) in studies of the mucilage sheath with which the former genera are invested, and for related observations of the cell walls. My technique for sessile protozoa (eg *Vorticella*) is described on p186. For the study of fresh material at the present time, the slide and coverslip obviate the need for such preparations as Leeuwenhoek needed. Indeed, it is likely that the diffraction fringe which he must have found of great value in delineating the outline of the colony would only serve to detract from clear viewing under today's lenses. However, it shows clearly that Leeuwenhoek was accustomed to look for new methods and, after being given the stimulus of earlier published techniques, would move on to find his own.

He prepared smears on films of mica, too, and imprisoned drops of water between two mica slips for certain observations. This technique is a close approximation to the slide and coverslip mount used today, of course. But for his studies on bacteria he preferred to use specimens in water, mounted in fine glass tubes fixed with glue or wax to the point of the needle on one of his simple magnifiers.

Though the design of his microscopes was essentially simple —ie single-lensed—it is possible that Leeuwenhoek carried out his observations by holding an eyepiece lens near to his eye, a postulation I have described elsewhere. But the single lens was undeniably used by him for his countless routine inspections of specimens at a lower magnification, and these are the ones he always showed to visitors.

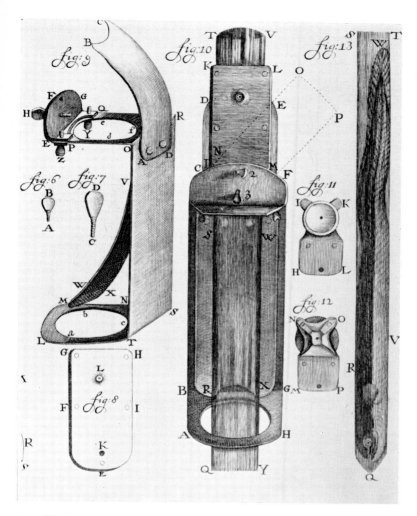

Page 33 A rare engraving of Leeuwenhoek's microscopical methods, showing the typical microscope (fig 8) mounted on a frame (fig 9) and held in position by screw H. A small fish (such as an elver) is dropped into a water-filled tube (fig 13) and this tube, by being positioned in the holder as shown in fig 10, enables the observer to see capillary flow in the tail membranes (fig 13, W). The microscope, as seen in fig 10, can be moved to one side (OP) enabling the specimen to be moved into the correct position

Page 34 (*top*) Drawing by Hooke (from *Micrographia*) showing his method of working. Note the oil lamp with glass reservoir (K), spherical flask acting as condenser lens (G), mechanical stage mechanism equipped with pointed styulus (M) for object, and screw focussing mechanism (G) operating through threaded collar. (*bottom*) Plate from Hooke's *Micrographia* showing a magnified drawing of woven cloth. This figure would have had a great attraction for Leeuwenhoek—a draper

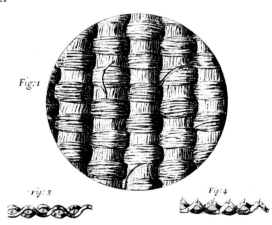

Leeuwenhoek's greatest contribution to microscopy was his emphasis of micro-organisms as distinct forms of life—a very different process of thought, I believe, from Hooke's magnification of already familiar objects. Hooke observed micro-organisms in infusions of seeds in water, in confirmation of the Dutchman's results, but he was no microbiologist.

Leeuwenhoek's interests in micro-organisms arose one afternoon in August 1674, when he passed by a lake he knew well, only to find it dotted with greenish, slimy tangles. The local folk believed this to be due to dew falling at the end of a summer day, and Leeuwenhoek took some home to examine with a microscope. He found it, under the lens, to be strewn with green spiral configurations 'after the manner of the copper or tin worms that distillers use to cool their liquors' and the size of 'a hair on one's head'. It was *Spirogyra*, of course; and the sight started him on a lifelong quest for more micro-organisms. He discovered spermatozoa, and, even more remarkable, he worked out what they were and where they came from. He observed bacteria and wrote about them with sufficient lucidity for us to be sure that it was bacteria he was seeing, and not some more robust organisms. He watched *Vorticella* dancing on its neat, spring-like spiral stalks; described rotifers and ciliate protozoa with clarity; and by diligence and devotion he came to be regarded as the founder of microbiology—a title (for all the exaggerated acclaim heaped on him in the past), which he richly deserves.

Yet some of his observations were erroneous. That is hardly surprising—he was, after all, breaking new ground. But what is unexpected is that, firstly, other students of his period have assiduously avoided drawing too much attention to the fact; and, secondly, some of the mistakes he made were of a very obvious nature even for the time. He recorded that red blood corpuscles were composed of six smaller spherical globules, for example; and even managed to see collections of distinct spheres in the heads of spermatozoa. This may have contributed to the so-called 'globulist' theories, in which every living structure was believed to consist of rounded sub-units. No doubt this was due in part to diffraction effects, which can—with simple lenses—give rise to a range of spurious optical appearances. But it was also due to a measure of wishful thinking on Leeuwenhoek's part, and no doubt involved a good measure of careless observational technique too.

By far the most surprising aspect of Leeuwenhoek's technique of microscopy, however, was the fact that he saw the specimen/plate/lens system as a single entity. He did not take up a microscope, and examine specimen after specimen with it (as we do today). Rather, he tended to make a microscope for a particular specimen and he kept it mounted permanently on the pinpoint in front of the lens. The microscope became part of the preparation. By the time he died he left behind several hundred separate assemblies, many of them with the specimen still in place. This is a unique attitude toward microscopy, similar to the magnifiers known as 'flea glasses' that were popular in the decades up to and around the middle of the seventeenth century, but quite different from latterday concepts in microscopical technique.

The large number of his microscopes that were extant at the time of his death (reputedly over 500, even though only about nine are still known to exist today) gave Leeuwenhoek another niche in history, for he was surely the first microscopist to go in for mass production. But Leeuwenhoek's microscopes were very hastily made (unlike his precision-ground lenses) and frankly crude in appearance; they were, moreover, intended for his own personal use and not designed for commercial distribution.

It was a fellow Dutchman, Samuel Musschenbroek, who began to make microscopes for other microscopists, and he and his son Johann became well established as instrument makers. The Musschenbroek microscope was a single-lensed device, but made with far more care and *finesse* than Leeuwenhoek ever revealed. It was possible to change specimens easily with the Musschenbroek microscope, and even to change lenses; and a small metallic quadrant bearing perforations of different sizes served to regulate the light entering the lens—certainly the first example of stopping-down similar to the modern use of an iris diaphragm.

But as the seventeenth century drew to a close, the growing interest in microscopy encouraged other designers to think about microscopes that were easier to use. The essential point was that a great deal of skill was needed to observe a specimen in any of the instruments then known. What was lacking—and what was going to be needed, if the subject was to lend itself to other fields of endeavour—was a foolproof form of microscope that anyone, a non-scientist for instance, could pick up and use like a telescope or a simple magnifying glass.

There were two problems: obtaining specimens suitably mounted, and the difficulties of manipulation and focussing. The answer was to find a compact and sturdy housing for the lens, and many workers in the last two decades of the century found their own answers to the problem. In 1685 an Italian named Tortona reportedly announced a form of instrument in which the focussing was in some way carried out by screwing the lens holder in or out; and other workers in Italy and elsewhere produced designs for compound microscopes which focussed in this manner.

In 1694 Nicholas Hartsoeker announced a microscope in which this principle had been applied to a simple, single-lensed magnifier, and in 1702 the British optical instrument maker, James Wilson, began to produce them in large numbers. Though he was only the populariser of the design, and not its innovator, it has been known as the Wilson screw-barrel ever since (see plate, p52). The microscope was simple to use, indeed examples I have handled and used are so efficient, portable and reliable that one is tempted to wonder why a similar device is not in production today. The Wilson was very cheap even then (it costs about two pounds) and with its kit of ready-mounted specimens it suddenly meant that microscopy could become a hobby for the casually interested amateur.

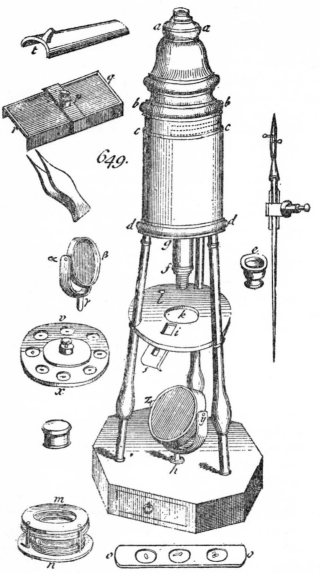

Fig 4 An early microscope of Culpepper-Scarlett form dated circa 1730. Note the three-specimen slider (o-o) and design for a wheel of specimens (v-x)

The objects were held in an ivory slider less than 1cm in width. Three or—more usually—four were situated in each separate slider, and they were introduced into the microscope by being inserted between two brass plates. The plates were circular and had an aperture in the middle; they were each inset with a locating groove that prevented them from being rotated but allowed them to slide apart easily enough, and they were held together by pressure from a strong spring. The position of these slider-retainers, as we may call them, was determined by screwing in or out a threaded element bearing at its 'distal' end a condenser lens. The magnifying lens itself was housed in the centre of a metal cap which could be easily unscrewed and changed for a unit containing a lens of different power. This was fitted to the end of the microscope body, at the opposite extremity to the condenser lens, and as the condenser unit was screwed home the slider-retainers (and hence the specimen between them) were gradually brought nearer to the lens until they were seen to be clearly in focus. It is evident from this mode of construction that the condenser lens was kept at a fixed distance from the specimen.

The whole device was only about an inch in diameter and a couple of inches long: and on to one side of the main body tube a handle of brass or ivory was screwed. The microscope could then be focussed, and passed from hand to hand so that successive observers could study the specimen, each individual peering through it as though looking with a telescope at the sky. A bright northern aspect was the chosen background for observations, though a candle or lantern flame would do as well—but, as the instructions warned, it was necessary to ensure that your wig didn't get in the way, or the quality of the image would have been impaired...

The kit of ready-mounted specimens consisted of a selection of ivory sliders roughly 6cm in length and 1cm wide. They were perforated with four apertures (three in the earlier versions) as shown in the figure. Each hole was countersunk and had a seat-ing shoulder halfway through the thickness of the ivory (each slider being roughly 2mm thick). Against this shoulder a disc of mica would rest. The specimen was placed in the centre of the disc with a pair of pointed forceps usually supplied with the kit, and a second mica disc was dropped on top of it (somewhat like adding a coverslip to a present-day dry mount). A thin brass circlip was then compressed and introduced into the cavity so that it held the mica slips tightly together.

The sliders were supplied in a slender container similar in construction to an old-fashioned open-razor box. The lids slid off at either end, and four or five sliders could be inserted (making eight or ten in all). The specimens that were supplied were not mounted in any particular order (eg botanical or zoological material on separate slides) but were in the form of an assortment of familiar objects. Typical examples would be:

TS pith of *Sambucus nigra* (elder)
TS pith of *Juncus* sp (stellate cells in the rush medulla)
*Pediculus humanus* entire (the louse)
Fern sporangia
Human hair clippings
Skin scale of *Solea solea* (the common sole)
*Funaria* apex (branch of common moss)
*Musca domestica*—eye showing lens structures

and so forth. No identification was given on the sliders. A small box containing spare mica slips and circlips was provided in addition, to facilitate the production of home-made specimens.

Many of these instruments were equipped with a glass tube in which small aquatic specimens could be examined, and there was a facility for the observation of capillary blood flow in the fins of a small fish. But the main technique of microscopy, as such, was simply the mounting of dry specimens between mica discs in an ivory slider.

This was, without question, one of the simplest and most foolproof of techniques. It is easier to obtain good results this way than it would have been to impale a whole insect on a pin. And for all its simplicity, it did confer clear advantages: it enabled the observer to utilise transmitted light, in itself a very important step in 'mass-production' microscopy; and the same components could be used time and time again to mount different specimens if need be.

The result was that microscopy entered a period of change and development almost overnight. Now anyone could own a microscope, not just the élite who were affluent enough to afford one. And—most important—it could be used to advantage for a range of purposes, quite unlike the earlier microscopes. The introduction of the slider (from which the term 'slide' derives) through the development of the drum microscopes that followed, used right up to the early decades of the nineteenth century, was —like the 35mm transparency in amateur photography—the key to a further growth in popularity.

CHAPTER THREE

# THE ERA OF BRASS AND GLASS

IT WAS about the middle of the eighteenth century that the microscope took a sudden step and became an instrument similar in its basic appearance to what we use today. The optical system was the same as it had ever been, but with the microscopes made by Cuff, the famous English craftsman, microscopy began to become a precise matter—and not just a question of guesswork and enthusiasm. Cuff's microscope tube was made of brass; a conventional adjustment for focussing was fitted to the main supporting structure of the stand; and the recognition of the system as sound was revealed when many other manufacturers borrowed the principle and produced similar designs themselves up to the turn of that century. One of the users of the basic principle was George Adams, who also fitted a disc bearing a range of objective lenses (from 3/4 inch to 1/20 inch in focus) which enabled them to be quickly changed. This we would recognise as the precursor of the modern rotating nose-piece. Adams had first used the idea of a 'wheel of lenses' on his New Universal Single microscope dating from 1746; but this had used a set of simple lenses. The adaptation of the principle to the objectives of a compound instrument was an important new aid for the microscopist.

At about the same time the first projection microscopes appeared. The idea was first noted down by Benjamin Martin in the early 1770s; his design was for a microscope in which the object was illuminated by a concentrated beam of sunlight and could be projected on to the wall of a darkened room. He called it an 'Opake Solar' microscope, and the description 'solar microscope' was applied to this type of instrument subsequently. By the turn of the century they had become popular playthings. The instrument—surely the nearest one could get to colour television at that time—was used to project a living louse or a flea on to a wall. George Adams wrote that it was 'stupendous' to see a flea 'augmented' to as much as 10ft in length, 'with all its colours, motions and animal functions distinctly and beautifully exhibited'. It was the sight of the projected image that encouraged some scientists to record the shadows on light-sensitive paper. An English scientist, J. B. Reade, is reported to have taken pictures in this way as far back as 1836. If this is so, then not only were these the first photomicrographs, but they have a good claim to have been the very first photographs as well.

In Britain, where the unreliable climate was unsuitable for the use of a solar microscope, an alternative was sought, and Adams produced the 'Lucernal Microscope'. It had a bright oil lamp as an illuminant, and would produce a brilliant image of the specimen if properly used.

For the serious microscopist, these instruments were only playthings—yet there was in them the seed of an important new advance that would help the technique tremendously. Martin, in order to produce a clear and true-to-life appearance of the image in his 'Opake Solar' microscope, had built a corrected achromatic lens for it. It was a triple-lens combination and gave a well corrected image. The view so often studied by the microscopist until this time, was blurred to a greater or lesser extent, subject to a great deal of distortion, and—most annoying to the eye—it showed ample signs of colour fringes due to the different powers of refraction of glass to light of differing wavelengths. By using a selection of positive and negative lenses Martin found it was possible to produce a microscope lens that was colour corrected to a pronounced degree. Achromatic lenses for telescopes were already well known, and Martin's lens was not adopted by manufacturers for other kinds of microscope as the practical problems were thought to be too great. His lens was used for projection, and did not have to be very powerful in terms of magnification. To adapt the same principle to a conventional objective lens would have been difficult.

But this was clearly to be the next step for the microscopist. The deficiencies in the image quality at that time were already producing some quaint results in the scientific uses of the instrument. Xavier Bichat, probably the true founder of histology as a discipline in its own right, often said that microscopists were being deluded by putting their own interpretations into the observed image, to a degree far above anything that could be justified by the lenses themselves. Many observers claimed to see globules or fibrils at the limits of resolution, which they imagined to be some form of the ultimate structure of matter; and talk of 'microscopical deception' was heard in learned circles.

But in the 1820s all this began to change. J. J. Lister was giving weighty considerations to the theory of achromatism, with the eventual aim of designing lenses mathematically, rather than by trial and error, and C. R. Goring, an Edinburgh doctor of medicine, made an important advance by the introduction of test objects. He had commissioned a microscope from a Scottish manufacturer named Aidie, but Goring was far from satisfied with the optical quality of the finished instrument. It was then that he began to look for objects that he could use to assess the power of the lens system to resolve detail. He realised that natural structures had a high degree of consistency, so that the diameter of the lens elements in the eye of a housefly, for example, would be consistent to a high degree no matter which individual fly of the species was examined. Goring soon settled for wing scales of butterflies as suitable test objects for routine use. They are marked with very fine grooves and ridges which are of a relatively constant size, and using these as his point of reference he began to look carefully through different lenses in order to gain a real idea of the image quality. It was his work that made microscopists realise how arbitrary their own results often were.

A glance at the dimensions of a human red blood cell (erythrocyte) will reveal the extent of this confusion. The diameter—actually about 7.2 µ$m$ in a dried smear, 7.5 µ$m$ in life—was variously given as 24.4 µ$m$ (Bauer, 1816) at the one extreme and 4 µ$m$ (by Young and Wollaston) at the other! Little wonder that the microscopist tended to trust his subjective instincts, rather than what he was told by others. With Goring's idea of a test object, it was at last possible to make some realistic trials and the comparison of image quality became feasible for the first time.

Of course, at these high magnifications even small discrepancies become important. It occurred to a German instrument manufacturer, Friedrich Nobert, that lines ruled on glass would make a calibrated and unarguable test object and in 1845 the first of the Nobert test plates appeared. He followed the techniques of Fraunhofer who had earlier produced diffraction gratings by ruling parallel lines on glass of the same order of frequency as the wavelength of visible light. Nobert's plates were ruled with a series of parallel lines in groups, each group calculatedly closer together than the last; and for the microscopist it was a simple matter to use such a plate and simply to note the number of the finest band in which the individual striations could be perceived.

Though Nobert did not know it at the time (the science of optical theory had yet to develop this far) many of the finer bands on his improved test plates were too fine to be resolved by any optical microscope. He had exceeded the limits that light could offer, and it was not until the electron microscope was brought into the picture and used to examine the finest bands of all—with separations as fine as 0.11 µ$m$—that it was first possible, in 1966, to see the precision of his results. None of his calibrations was more than 10 per cent away from the intended result, even for those bands that he would never live to see resolved.

By this time, J. J. Lister had completed his calculations on the mechanisms of refraction. They were published in 1829. From that time on, the development of microscopy took an entirely new turn. Equipped with the theoretical knowledge to design better lenses, and with test objects available, microscopists began to search for ever finer detail in their specimens and to be discerning about the lenses they used. Men devoted a large part of their lives to the resolution of the next band on a Nobert test-plate, as though it was a contest like an athletics meeting with record timings at stake. Fierce controversies raged over the relative merits of different lenses—and the manufacturers were spurred on as never before. The backlash and uneven construction of many of the components of older microscopes proved to be irksome as the optical quality of the lenses improved, and several manufacturers began to turn their attention to mechanical perfection—they became devoted to finer and ever finer standards of machining and consistency, and the finest age of microscope construction began to dawn.

The two names that stood out at this time (the 1830s) were Ross and Powell. Both originated firms whose standards of technical excellence are legendary to this day. One writer in 1841 wrote that the element of competition between the two companies was the secret of their success—a competition kept up 'by the incessant demands of microscopic investigators . . . and the results have been, and are daily manifesting themselves at their hands, are of the most important and satisfactory nature'.

The third of the great manufacturers was James Smith, and it was he who was commissioned to produce the first achromatic microscope for J. J. Lister; it was finished in May 1826. Smith was slow off the mark, however, and did not begin manufacture on a commercial scale until 1840 or thereabouts. The three names —Powell, Ross and Smith—came to mark the ultimate standards of technical skill and craftsmanship. A study of the designs of the three will reveal a tendency for successful developments by one to be plagiarised, and then improved, by another; and certainly many of Powell's designs seem to have borrowed heavily from Ross. Of the three, it was Hugh Powell (eventually in partnership with his brother-in-law, R. H. Lealand) who had the edge on the others. No one has matched his standards since.

Powell was born in 1799. He trained as an instrument maker and at the age of 30 he was approached by an inventor named Cornelius Varley, 18 years his senior, and requested to construct a microscope according to Varley's personal requirements. Powell brought a new insight into the field: he had not been indoctrinated by the orthodoxy of microscope manufacture at the time and he worked on the design of Varley's stand with diligence. At once he realised that one of the microscopist's oldest enemies —looseness or backlash in the controls—could be eliminated by incorporating a spring in such a position as to maintain the bearing surfaces in close contact at all times. It was something of a breakthrough in design. To turn one of those controls even today is to experience precision of an unsurpassed level.

Varley was one of the founder members of what was to become the Royal Microscopical Society, and in 1841 the Microscopical Society of London—as it then was—commissioned Powell, Smith and Ross to each supply a microscope for assessment according to 'their own peculiar views'. It was in the same year that Powell joined forces with his brother-in-law, P. H. Lealand, who had been assisting with the production of lenses for several years previously.

The union brought the meeting of two brilliant minds. Indeed the stands that were being manufactured in the early 1840s had no fine focussing adjustment: the smoothness and precision of the high-ratio 'course' adjustment was sufficient for all normal purposes. No Powell and Lealand microscopes were displayed at the Great Exhibition in 1851, where Andrew Ross dominated the picture. It has been suggested that Ross amounted to opposition of too formidable a nature—but within eight years he was dead. In 1861 Powell and Lealand announced their Large Compound Microscope stand, which was well received. Its controls were fine, its manufacture as near perfection as could be conceived. The stand was fitted with a rotating stage in which the substage also rotated (thereby remaining in perfect alignment), and mechanical stage controls were fitted too. The stand, at that time, cost over £35, the optical accessories adding an extra £50 or so to the price. Even so, it was a popular microscope, and it gave rise to the legendary Powell & Lealand No 1, probably the finest instrument in the history of microscopy.

The substage was now fixed, so that it did not rotate with the stage. The centering screws for the substage mechanism were carefully designed for ease of operation, and even on century-old instruments they perform with something approaching grace—a kind of silky subtlety of movement like turning a drinking-straw in a bowl of syrup. Hardly any modifications were made to this model as the years of production passed. In 1882 a fine adjustment was added to the substage—the first time it had been done, and one of the last—then in 1887 a rack and pinion adjustment was added to the draw-tube as an aid to alterations of length with objectives of different corrections. Finally, a diagonal rack was substituted for the conventional right-angled rack-and-pinion thread in 1897. And so the 'No 1' remained in continuous production until the twentieth century. Pressures of a developing economy caused a gradual drop in output and in sales—the time taken on each stand was impractical (it has been recorded that one worker would spend as much as a whole day to perfect the helical screw thread controlling the lateral movements of the stage) and the last No 1 was made by the firm in 1901, when it had been reduced to just two men. Some more were made by an ex-member of the team, which were released for sale after being approved by Thomas Powell, but that was the end of the production story. Each of the instruments was individually checked and signed 'Powell & Lealand', and there had never been more than half a dozen workers in the firm at one time. And so the great era of microscope design came and went; and with it died a kind of altruistic craftsmanship that seems inconceivable to us today.

The effect on microscopy had been profound. The mechanical deficiencies of older stands had been only a slight drawback, since they did not interfere with the results that could be obtained from generally inferior lenses. But the microscope at the end of the nineteenth century was a very different matter. A vast range of accessories was available, and the quality of the lenses was higher than anything available from a modern manufacturer. The work of the microscopist had become largely automated. By a turn of a carefully constructed screw mechanism he could adjust the substage, change objectives, delicately manoeuvre the microscope slide; it was the difference between a hand-finished limousine and a rattle-bones horseless carriage.

The optical equipment underwent startling changes following the publication of Lister's work on achromatism. It was Lister (father of the famous surgeon who introduced antiseptics into the operating theatre) who produced the mathematics—though he was not the first to make effectively achromatic lenses. Quite apart from Benjamin Martin's exceptional lens (as we have seen, not truly a microscope lens at all) dating from 1774, a little-known optician named Harmanus van Deijl who worked with his father, reportedly made an achromatic objective at about the same time. They put their lens into production in the early 1800s. In 1791, an amateur Dutch enthusiast named Beeldsnyer announced that he had made an achromatic lens which was never produced commercially. Another amateur, this time an Italian named Marzoli, put an achromatic lens into production between 1808 and about 1811. Even when Lister came on the scene, with his hard-and-fast calculations, opticians were still not ready to listen. He set out to learn the techniques of lens grinding and polishing, so that he could carry out the work for himself. In 1826 he ordered a special microscope from W. Tulley which was to carry these achromatic lenses based on the formula. The microscope is believed to be the very first which had been built for use solely as a compound microscope. All the earlier models, from those made by Cock for Hooke a century and a half earlier, had been intended

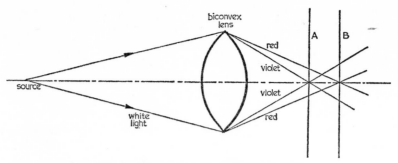

Fig 5  Diagram of chromatic aberration

for occasional use as single-lens simple microscopes when the greatest capacity for resolution was required. But Lister was so confident that his corrections would give the best possible image (even if not, as one recent researcher has over-confidently claimed, 'entirely free from both chromatic and spherical aberration') that the use of the single-lens would be superseded.

Probably the last of the great discoveries made with the single lens were those of Robert Brown. He observed the agitation of suspended small particles, due to molecular bombardment, and still the most attractive demonstration of molecular and atomic movement, known as Brownian motion to this day. And in 1833, as a vital clue in the formation of the 'cell theory' of the structures of living organisms, he identified and studied the nucleus of plant cells, as well as making numerous profound and lucid excursions into the wider realms of plant anatomy.

But Brown was the exception. Lister's microscope stand had introduced the idea of body, stage and mirror being mounted in line as a unit which could be inclined back on a heavy metallic base (or foot)—the design of choice in the 'conventional' microscope of the first half of the twentieth century, which was superseded only when the inclined eyepiece attachment gained popularity in the 1960s. The orthodox high power lens of the 1840s had a resolution of only about one micron, which is far worse than the performance of a student microscope today and is

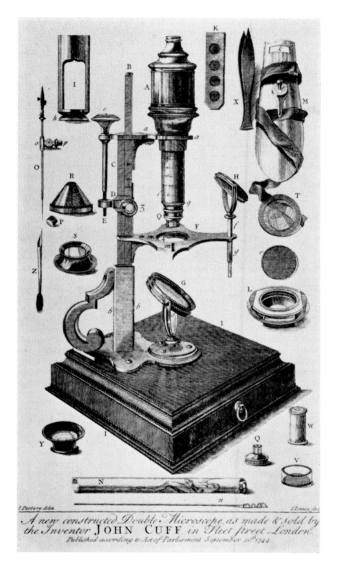

Page 51  A very fine engraving by Tinney of Cuff's 'Double Microscope' dating from 1744. The tapes around plate M were used to hold firm a frog so that the circulation could be viewed through aperture *k*. Note the drawer in the base wherein accessories were stored

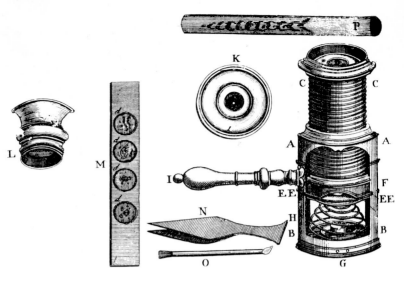

Page 52 (*top*) Contemporary engraving of the Wilson screw-barrel ready for use (circa 1720). The specimens are mounted between mica slips in the slider M. This may be slid into the microscope between the brass plates EE-EE. The small single lens, mounted in plate K, is screwed into the end of the microscope at G and it is through here that the observer looks. Lens C-C is the condenser. (*bottom*) Microscopy a century ago: a picture of the Powell & Lealand No 1 stand in use. Note the high-intensity oil lamp (right) and the shaded lamp for the drawing-pad (left). This conception owes something to artistic licence; as it is, the hot exhaust fumes from the left-hand lamp would go straight up the microscopist's nose

even equalled by many 'toy' microscopes made for schoolchildren. From the 1/6 inch lens of that period to the 1/80 inch lens made by Lealand in 1872, the progress was rapid. About 1840, Ross first produced an achromatic objective with a 1/8 inch working distance and with a calibrated collar which could be adjusted to suit different thicknesses of cover-slip. Each of these lenses was signed with the maker's surname, and the style of writing (with the first 's' written in script like an 'f') lends an air of antiquity to a remarkably advanced lens design. The compound substage condenser appeared, too; or rather reappeared. It had been described by Bonanni in his horizontal microscope of 1691, and single-lens condensers were fitted to many earlier models (including the screw-barrel microscopes that had been so popular). But the demands of steadily improving objective lenses meant that the uncertain, yellowish light of a gas or oil lamp needed to be concentrated, and daylight needed control; and so the condenser appeared as a double lens unit. We now call this form, as a rule, an Abbé condenser; but Abbé was not to appear on the scene for a few years yet, and his main role in connection with this particular concept was to popularise it, and not to be the innovator at all.

After the death of Ross, who did more than anyone else to introduce and usefully exploit the findings of Lister's calculations, Powell & Lealand continued to press ahead with their own version of the achromatic objective. In the year of Ross's death, 1859, the shortest focal length lens made by Powell & Lealand was 1/16 inch. By 1868 they had produced a 1/25 inch lens; in the same year they announced a 1/50 inch version; and in 1872 they made their unique lens at 1/80 inch. Yet microscopy was not only benefitting from better objective lenses—there had been another dramatic development which altered, once again, the whole direction of progress.

This was the realisation that lenses could benefit greatly from being optically joined to the preparation by means of a liquid. At first this was seen simply as a way of avoiding light losses through scatter and reflection. Hooke had tried this out, it will be recalled, and in the Paris Exhibition of 1855 an Italian named Giovan Battista Amici—a remarkable microscopist, to whose reflecting microscope we shall later refer—demonstrated an objective that used water as the immersion liquid. In the 1860s, immersion lenses were generally available, but their performance was generally inferior to that of equivalent dry objectives and (as their main benefit was believed to be a brighter image, with lower light loss) they did not gain widespread popularity.

None the less, the efforts of Edmund Hartnack—an English optician who made the first immersion objectives on a 'production line' basis—should not go unnoticed. It was through his efforts that the immersion principle was promulgated, and microscopists throughout the Western world came to hear about it. Even if the principle was still only a curiosity that the intelligent microscopist should know about, if not use, its very existence served to prepare specialist opinion for the breakthrough that was to come. In 1869, Powell & Lealand announced an immersion objective which was found to excel in resolution. It was capable of resolving the band on Nobert's test plates in which the lines were separated by a mere 0.225 $\mu m$, and was hailed as being the best available at that time.

The resistance to the idea showed by many microscopists, as opposed to opticians and theorists, is exemplified by the writings of a very able microscopical investigator, Francis Wenham. Many contemporaries believed him to be the best and most knowledgeable microscopist of his time. He was, indeed, a consultant advisor to Ross. But he scorned the immersion principle bitterly and went on record as saying that instead of applying a layer of fluid to the front of the lens, thereby increasing its thickness, one might as well simply manufacture thicker lenses in the first place and have done with it.

The end of this attitude was already heralded in the 1860s in the German town of Jena. There, Ernst Abbé—son of a textile spinner—had graduated with honours at the local university and was appointed a lecturer in the mathematical astronomy department at the age of 23. In 1866 he was invited to join the 20 year-old Jena firm of Carl Zeiss as optical consultant, and he set to work to design the first successful series of immersion lenses in the history of microscopy.

Abbé showed that the ability of the observer to see fine periodic structures (such as the pattern of markings on a diatom, for instance) relied on the presence of light rays that were often diffracted at such an angle as to bring them outside the area of light collected by the lens. Abbé realised that the only way in which these diffracted elements of the light beam could be used in forming the image, was to make the lens larger—to increase its aperture, in other words. It was this concept that proved to be a turning-point in lens design, and Abbé's understanding of the significance of aperture (quite apart from magnification or focal length) led to the coining of his term 'numerical aperture'. This figure was defined as:

$$N.A. = \sin a$$

for a dry lens, where $a$ = half the angle that the lens subtends from the object in focus. Since that time, other workers have shown that the important parameter is the ability to define as entities two points (assumed to be infinitesimally small) separated by a distance D. Various interpretations of this relationship have been given:

$$D = \frac{\lambda}{2 \, N.A.} \quad \text{-------- (1)}$$

$$D = \frac{0.61 \, \lambda}{N.A.} \quad \text{-------- (2)}$$

$$D = \frac{\lambda}{N.A._i + N.A._o} \quad \text{-------- (3)}$$

$$\frac{\lambda}{N.A.} \geqslant D \geqslant \frac{\lambda}{2 \, N.A.} \quad \text{-------- (4)}$$

$$D = \frac{1.22\,C}{\text{N.A.}} \qquad \text{-------} \quad (5)$$

where: D = separation (distance) of particles
$\lambda$ = wavelength of illuminant
N.A. = numerical aperture
(N.A.$_o$ = N.A. of objective system;
N.A.$_i$ = N.A. of condenser)
C = a constant between 0.4 and 1.0 which relates to the assumed size of an 'infinitesimal' point.

But these equations—interesting as they are—should be regarded more as mathematical gymnastics than the stuff that microscopy is made of! They relate to the nature of the particles, different ways of defining the distance between them, and certain refinements of diffraction theory that are of academic interest above all. The approach in Abbé's formula meant that microscopists had a universal parameter by which to compare lenses. If the N.A. of one lens is exactly twice that of another, so is its resolving power. And that meant a great deal in days when initial magnification (much of it empty and superfluous, for it did not reveal extra detail) seemed to matter above all else.

In 1878, an English microscopist, J. W. Stephenson, wrote to Abbé with a new proposal. He put forward the concept of homogeneous immersion—a system in which the refractive index of the immersion medium was the same as that of the glass in the lens element. It acted in this way as a means of optically joining the objective to the coverslip, not just as three conjoined elements in the system, but as a single unit. Abbé wrote that the idea had occurred to him earlier, but he had put it on one side as being impractical. With renewed interest, the University of Jena set out to find a liquid of suitable refractive index and settled for cedar-wood oil. This was adopted as the immersion liquid of choice and has been widely used since, until the production of synthetic oils began in earnest after World War II. But was

Stephenson the individual responsible for the idea? Perhaps not. In 1873 an American microscopist, Robert Tolles, had designed and built a homogeneous objective which used thin Canada balsam as the immersion liquid. Tolles and his work were known about in England, and it is possible that it was this earlier development that acted as the trigger to the idea in Stephenson's mind.

The Powell & Lealand lenses that followed came close to the theoretical limits for immersion systems. Soon they had in production a 1.43 N.A. objective of 1/12 inch focal length and this was followed by an objective with an N.A. of 1.50. As it happens, 1.52 is the maximum possible in such a system, and so this must be seen as an extraordinarily powerful lens. It is said that Powell and Lealand themselves would not allow anyone to see them at work in the optical department of their firm, and the work in progress was covered up by sheets should anyone venture in. Their lens mounts were made so that, once assembled, they could not be taken apart; and so only a little is known even today about the details of their construction.

Yet still there was one further step to be taken. The achromatic design of Abbé's lenses was excellent, but it contained one inherent drawback. The construction was based on the use of soda- and flint-glass elements that could, in combination, bring to a common focus the two ends of the visible spectrum. The yellow-green light in the middle of the spectrum, however, was not focussed with complete efficiency, and an astute observer using a high-power eyepiece would have noticed a fine yellow fringe in certain specimens. It was certainly not a drawback to the majority of users, nor did it hinder the bulk of the serious work under way at that time; but Abbé—like any enthusiast—was determined to eliminate even this fault if he could.

His answer was to search for different kinds of glass, and eventually a fellow-German named Schott came up with a lithium-glass that seemed to provide the qualities Abbé wanted. The first lenses that the Zeiss company produced in 1886 (Abbé was now the head of the firm) were optically advanced, but after a few months of use they began to become opaque and eventually were unusable due to a failure in the new, still experimental glass. But by the end of the century these problems had been overcome and many other companies (including Powell & Lealand, who marketed one of these lenses with an N.A. of 1.50 in 1893) were producing versions of their own. Abbé did not patent his idea for apochromatic lenses, as he dubbed them, and the range of models that appeared was wide. It is interesting to note that Abbé went on to use fluorite—a naturally-occurring transparent mineral—as a component of his later apochromats. Today 'fluorite lenses' are manufactured by several makers and are better than the conventional oil-immersion lenses made from glass. They provide a useful compromise between orthodox lenses and the apochromat.

Personally the best lens I have seen manufactured in the past few decades rates at an N.A. of 1.4. Most equipment will include an oil-immersion objective lens with an N.A. of 1.30, no more; if a 1.50 N.A. lens is required then I fear your only opportunity to acquire one would be to scour the second-hand shops and antique dealers for a dusty little box that no one has looked in for fifty years or more. With luck you might even come across a rarity such as the Zeiss objective made to use naphthalene monobromide as the immersion liquid. That lens (dating from 1891) was stated to have a numerical aperture of 1.60, and that is about as far as anyone could go.

The availability of microscopes and lenses such as these was, of course, a direct stimulus to microscopy as a technique in its own right. Yet the uses to which the new facilities were put often took a very long time to materialise. Clearly, a certain refinement of microscopy at the technological level did not necessarily indicate a comparable step in scientific discovery.

Exactly 200 years after Hooke's first experiments, in 1863, Henry Sorby examined the smooth surface of steel under the microscope and so laid the foundations of metallurgy. The structures he was looking at could, in retrospect, have been seen by Hooke had he ever thought to look for it, instead of being content to search for the scratches on the razor's ground edge that he expected to find. It was a century after Leeuwenhoek was examining bacteria before Muller drew several of them in his studies of 'infusoria', and another hundred years or more before the systematic study of bacteria was under way. And it was the best part of two centuries before Hooke's observations on the cellular structure of plant tissues was elaborated into the general concept that living things are, in the main, composed of cells as fundamental units.

And behind the whole trend was the development of microtechnique. At the beginning of the century the microscopist had very few options when it came to the preparation of a specimen. He either left the specimen whole, or he cut a thick slice of it with a pen-knife. Once he had done that, his most usual mounting procedures would be either to hold the object on the end of a pin (or in a little holder like the end of a pair of forceps), or to sandwich it between two circular slips of mica in a holder carved from ivory. There were, as a rule, four apertures—with a specimen in each—in these ivory sliders.

The earliest sliders were only about two inches long and less than half an inch wide, with each aperture being roughly $\frac{1}{4}$ inch in diameter. By the beginning of the nineteenth century a good many of them were about 3 x 1 inch, which was adopted as the best size to cut glass slips when these become more generally available. Glass was too brittle to be cut much smaller without risk of breakage. Around 1820 some microscopists were preferring to use glass slips of this sort rather than the delicate mica discs in their ivory holders. The object might be laid on its glass slider and covered with a disc of mica before being sealed over with a strip of gummed paper bearing a hole in the middle where the specimen was. By 1840 thin glass was widely available, and coverslips began to appear. The glass slider—now known increasingly as a 'slide' instead—became more widespread and has, of course, remained with us since.

The use of a mountant, joining the coverslip and slide together with the specimen in between them, became widespread at this time. It is fortunate indeed that the choice of this natural resinous material (at first used 'raw', later dissolved or thinned with ether or, as is used today, xylene) gave microscopists a cement that was very close indeed in refractive index to glass. Canada balsam was being written about in the 1860s as an imperfect mountant, but the best then available; yet it has remained with us since and is still found in microscope laboratories throughout the world. The synthetic materials that are now available have been slow to replace the traditional resin. There are records that the early telescope makers used Canada balsam to cement lenses together, and the first known use of the material as a microscope slide mountant was in 1794 when Ypelaar used it to secure mica slips on each side of the specimen.

But there was a totally different form of mounting in vogue at that time, which we seem to have forgotten about today. That was the use of ivory containers about 15mm in diameter and a few millimeters thick. Each one had a white or blackened base, to give contrast to the specimen mounted on it, and had a transparent slide of mica or perhaps thin glass. It was a popular form of mount in Europe from the 1760s onwards, though it never became widely used in Britain, where ivory sliders were favoured. The use of built-up varnish rings or circular cells on the standard 3 x 1 inch slide which was popular in the decades prior to World War I was similar in principle to this circular container idea, and the conceptual descendant of the same technique is the use of small square plastic boxes to hold mineral specimens too bulky to fit on to a slide as such (p158).

But by the middle of the nineteenth century, the 3 x 1 inch slide was universally used for all routine work; chambers or wider slides (3 x $1\frac{1}{2}$ inches, for instance) were used then as they are now for special purposes, but the glass slide has an enduring popularity. Even so, the late-nineteenth century preparations were usually covered with paper, often itself decorated with an engraved design, so that they still had much of the appearance of an old-fashioned 'slider' with but one cell. It was in the closing decades of that century that the slide with its labels at each end became increasingly popular throughout Europe. Today there are variations (such as the use of thick card labels, serving to keep slides separated during storage in drawers in laboratory collections, for instance), but basically the technique for mounting has been virtually unchanged from that time to this.

After the 1820s the growing use of glass slides meant that better micro-dissection was possible under the lens. It was now feasible to macerate a specimen with needles in order to tease out the various component parts and so gain an idea of the organisation of a tissue. At the same time, the idea of embedding tissues for sectioning began to appear. There was at first none of the sophistication of dehydration, clearing and embedding in the sequence we know today, of course; then it was basically a question of using the embedding material (candle wax or india-rubber gum) as a means of external support for the entire specimen, so that the sections could be cut with a penknife or a scalpel. Even a century ago the state of microtechnique was far less advanced than we tend to assume was the case. In 1871, just when the popularity of microscopy was at its height, in relaxation as much as anything else, an amateur naturalist living in Kent, John Martin, published a description in some detail of the state of microscopy. The techniques give an interesting insight into the primitive approach still in vogue:

> We will proceed to the preparation and mounting of objects, so as to give the reader insight into the matter.
> There are three forms of mounting chiefly used, viz in balsam, dry, and in fluid or semifluid. All these methods should be tried upon the matter that may come into the student's hands, as the full structure and beauty is often lost from the specimen being mounted in an imperfect manner. Canada balsam, although it has its faults, appears to be at present the most reliable substance in which the majority of transparent objects may be mounted. Most insects and the parts of the same are mounted in balsam;

they must be placed in the potassa solution from six hours (or less) to as many days, according to the softness or hardness, or transparency or opaqueness, of the object; experience will give the time. They must then be put under a slight pressure so as to squeeze out part of the extraneous matter; replace in the potassa solution for a short time, then put them under increased pressure until cleared of the rest of the contents, when they must be placed in a large quantity of warm water for a few hours. It is best not to touch the object at this stage even with a camel-hair brush; but the vessel of water must be repeatedly shaken and extra water added, so as to thoroughly clean the specimen from the potassa (this is important); it must then be taken out of the water and dried between two slips of glass. All the pressure requisite for a small object is obtained by the use of the American paper-clips; but if the object is large and thick (as, e.g., many of the beetles), the regulated pressure of a small screw-press is necessary. After it has been well dried (which, of course, will take from a few hours to as many days, according to the size and nature of the object) it must be soaked in turpentine, or, what is better, distilled Canada balsam, until moderately transparent; if small, it must not be taken from the glass slide. A slip of glass of the recognised size, 3 x 1 inches, must be taken, a drop of balsam (the size of the drop proportionate to the size of the object) placed with the glass tube in the centre of the slide; the drop must then be made to spread slightly by the use of moderate heat, the object placed in it, and the thin glass cover applied with care. If any air-bubbles appear, they will generally be found to have dispersed after a day or two; and unless the object is valuable, it is best never to attempt to remount it. As the patience of the learner would be greatly tried during the process, it is always better to begin again with a fresh specimen. The balsam takes a long time to dry if left to itself, which is best; but if the objects are wanted early, they may be dried over a gas-jet, or by any other plan, so that the heat be about 50°C; this temperature will not do for all objects. The next thing is to finish the slide neatly. Paper covers are sold for this purpose; but the best plan for durability is simply to finish with a ring of Berlin black varnish; or even this may be omitted in some cases; the advantage of this plan is that the slides can be kept much cleaner &c. The white paper label can then be affixed with the English and scientific names, also what fluid &c it is mounted in, and the date of the preparation. It can then be stored, if money is an object, in the cheap rack-boxes sold by Mr. Wheeler and other opticians at prices varying from sixpence upwards (these boxes are not covered with cloth); the boxes may then be numbered and placed on shelves in the same manner as books, whereby the objects are kept in a horizontal position.

Fluids had, until lately, greatly gone down in the estimation of microscopists as vehicles in which to mount various specimens; but glycerine and the more recent, if not so useful, substance silicate of potassa, which appears to be a very favourable semifluid for mounting certain structures, have again caused them to be used; and there are now varnishes and cements, such as the india-rubber and shellac cement, which, with care, will hermetically seal fluid preparations for many years. We will treat of the process. Take a slide, centre it on the turn-table, charge a camel's hair brush with the india-rubber cement, place the table in action and a ring or circular cell of the cement is formed, varying in depth according to the thickness of the fluid and the quantity used; turn a number of these cells and put aside to dry, to be used as wanted. With a pipette take from the alcoholic and camphor preservative fluid a sufficient quantity to fill the cell, soak the object in proof spirit for an hour or so to exhaust the air (it is better if the object has been kept in alcohol) or it may be done in much less time under the air-pump; it is then placed in the centre of the cell, the thin glass cover placed gently over it, so as to exclude all air-bubbles; soak up all the surplus fluid with blotting-paper, centre the slide again on the turn-table, and seal the cell with a ring of the liquid india-rubber cement; dry and finish with Berlin black varnish, then label as usual. Whatever fluid is used, the process is nearly the same; but when glycerine or chloride of calcium is used the cell must be sealed either with Bell's cement or a saturated solution of gum-dammar in benzole; when the silicate of potassa is used it is hardly necessary to seal the cell at all, as the fluid dries at the edges and seals itself.

The next form of mounting is the dry system. A cell made of cardboard, gutta percha, ebonite, &c, is cemented to the centre of a glass slip, sufficient Berlin black is then used to cover the bottom of the cell, a small drop of pure gum or any good cement is placed in the centre; the object, which has also a minute quantity of the same cement on it, is then fixed exactly on the same spot, and the slide is left under a bell-glass to dry, after which a circle or square of thin glass is closely cemented to the top of the cell; the slide may then be finished with any black varnish, or it may be covered with any of the paper covers. If the specimen is to form a transparent object, the Berlin black must be left out of the cell; and it is best not to cement the object to the glass slide, as the cement often shows through and spoils the appearance; the object may be fixed by the slight pressure of a thin glass cover in a shallow cell.

In the mounting of objects, great care must be taken to show the structural characteristics of the specimen; for if this is attended to, a greater amount of valuable information will be obtained even from a 'common object'.

*Algae, Confervoid &c.* These show well when mounted in a preservative fluid consisting of 1 part alcohol to 7 water, mixed with an equal quantity of a cold saturated solution of camphor in distilled water. There are some seaweeds with their fructification that show best when mounted in balsam; but if glycerine or any other fluid which causes a strong exosmotic action on the cell-wall be used as the preservative fluid, it must be done by its gradual addition to the water in which the Algae are contained, so that its action on the cell-wall and protoplasm will not be so abrupt as to cause any rupture of the same. A solution of alum is also often used in the preservation of some of the Algae: a fluid still better is the acetate of alumina, prepared by dissolving alum in acetic acid, crystallizing by evaporation, and to a saturated solution of this salt adding four or five parts of distilled water; or the acetate of alumina may be dissolved in the glycerine.

*Bone, Teeth, &c.* Sections of these substances, if required to be mounted dry, are best made by cutting a thin section with the fine saw, and finishing by grinding down with a file until they are made as transparent as possible; they may then be mounted in a dry cell. The sections are best ground by fixing them to a slip of glass with strong balsam; the better methods of preparation are: to cut a thin section after maceration in hydrochloric acid diluted with two parts water, then to mount in a cell with a fluid composed of acetic acid 1 part, water 2 parts, or the broken bone, tooth, &c may be placed in a fluid of 1 part glycerine, 1 part water, for a few hours, then add gradually a mixture of glycerine and acetic acid equal parts; after a short time thin sections may be cut with a fine scalpel. Mount in a preservation fluid, acetic acid 1 part, water 3 parts; or it may be mounted in glycerine, or glycerine and acetic acid equal parts.

*Crystals.* The formation of crystals, from saline and other solutions, under the microscope yields an extremely interesting and useful study, for example, the beautiful appearance of the crystals of chloride of ammonium caused by holding for a few seconds a glass slide that has had one drop of hydrochloric acid spread over the surface, over the fumes or vapour of ammonia: upon the gradual evaporation of this thin film of fluid, fine feathery crystals are formed; they are produced by the ammonia combining with the hydrochloric acid. These crystals may often be developed from the human breath, more especially in certain forms of disease.

Some forms of crystals are best produced by placing a drop of the solution under a thin glass cover and letting the fluid evaporate gradually. The majority of crystals formed from the various salts &c are best mounted in castor-oil and sealed with the gum-dammar cement; many show well when mounted in a solution of balsam in chloroform.

*Desmidieae &c.* For the mounting of these lower forms of vegetable life, see Algae; they are best collected by taking the green scum from the margins of ponds situated in open and exposed districts, and placing this green matter in a white saucer nearly full of water: shade, all but an inch or so, from the surrounding daylight; and in the space of a few hours, if fresh, the desmids will be found massed at the place that has been left exposed to the light; with a pipette they may then be separated from the surrounding substances, and mounted in a shallow cell with one of the preservative fluids. They are found in the greatest quantity of the later summer and the autumn months.

*Diatoms* are collected in nearly the same manner as Desmids, from which they may be distinguished by their light brown colour; they are often found growing in tufts upon the marine and the freshwater Algae. Their mode of preparation is rather difficult; but, in a few words, the following will be found the best process: burn the deposit in a platinum spoon until it assumes the appearance of a white ash, then boil in nitric acid for a short time, when most of the siliceous valves will be found quite clean: the large glass tubes used by chemists for collecting hydrogen and other gases will be found, on account of their length, of great assistance in separating the species according to their specific gravities; they may then be mounted dry, in balsam, or in some fluids, but not silicate of potassa.

The name of the species of Diatom, if known, must be immediately written on the slide. This rule holds good with all specimens, as, if a note is not made at the time, it is liable to be forgotten.

*Entozoa.* Many of this class of animals exhibit their anatomy best when mounted in glycerine; they also mount well, after preservation, in balsam. If surrounded by germinal matter, the use of the carmine or other dye to be used as a stain will cause the parasite to appear better, the dye staining the surrounding mass and leaving the animal untouched, an interesting species for observation will be found in the *Anguillula glutinis*, found in sour paste.

*Ferns and Mosses.* The investigation into the structure of the minute reproductive organs of the plants form an interesting branch of microscopy. Mosses (more especially at the time of the year when nature partially hides her glory) may be taken up as a special object of study; for their structural peculiarities are developed chiefly in winter. Many of the smaller species may be mounted entire, after soaking for a short time in water and draining off the same: they show well when mounted in the silicate of potassa; but most of the minute characters show best in balsam. The parts of the fronds of various species of Ferns which exhibit the sori show best when mounted dry.

*Lichens,* when entire, are mounted dry; but, to show the apothecia &c well, the section of the thallus must be mounted in glycerine or balsam.

*Fungi.* Sections of spores &c are generally best when mounted in the preservative fluids, as recommended for the Algae; but many of the micro-Fungi may be mounted *in situ* in a dry opaque cell; and some of the brands &c show best when mounted in balsam.

*Leaves and Petals.* The cuticles of these parts of plants form a large range for investigations. Most cuticles are prepared by boiling the leaf in a fluid made by adding about four parts of water to one of nitric acid; but the proportion must vary according to the nature and strength of the leaf. After the cuticle is separated by boiling in this fluid, it must be floated off from the waste tissue, delicately washed with a fine camel's-hair pencil, and mounted in a suitable fluid according to its thickness &c— if thick, in glycerine or balsam; if thin, in any of the fluids recommended for the Algae &c. The cuticles of the petals are best when torn from the surface; but for petals I prefer the colouring-matter to be nearly obliterated by the use of ether; then add weak sulphuric acid, dry, and mount in balsam; or some show best when mounted dry.

Most of the other vegetable tissues, such as spiral-vascular, scalariform, &c, are best mounted in glycerine, &c.

*Starches.* Many of these may be mounted in silicate of potassa, care being taken to moisten the starch first, or air-bubbles will be formed, which are difficult to get rid of in this substance without the use of the air-pump; if required for the polariscope, balsam is best.

*Insects,* parts of, &c, are best when mounted in balsam, although some of the smaller ones perhaps exhibit their structure better when mounted in acetic acid, 1 part acid to 2 parts water; they may be mounted in one of Pumphrey's ebonite cells, or in a cell made of the india-rubber cement; in both cases this is the substance with which to seal the cell.

*Palates,* or tongues of the Gasteropoda, a class belonging to the

Mollusks, are generally dissected from the animal cleaned with potassa, washed, dried, and mounted in balsam; they are then generally seen under the polariscope. Some, like the whelk's tongue, require to be slit up the centre, spread out and dried; they show well when mounted in glycerine.

*Zoophytes, Rotaria, &c* show best when mounted in a fluid as nearly as possible like their native element.

White slabs to be used for dissections &c are made of the white gutta-percha enamel (sold as a tooth-stopping) mixed with white wax; after this substance is run out into slabs about the 1/16 of an inch in thickness, they may be cut up and used at the bottom of the cells when it is required to exhibit any particular dissection in its natural position; for rough purposes ordinary gutta percha may be used, mixed with wax in the same manner. For dissections under water the gutta-percha and wax substance must be melted at the bottom of a deep white vessel; the porcelain dishes that photographers use will do for this purpose; a common dish, if deep, may be used.

In staining tissues the germinal or growing matter is coloured, and the formed or mature matter is not. But then, do not use Judson's dyes; they dye every thing. The addition of weak hydrochloric or nitric acid is useful for breaking up cellular tissue &c.

Thin sections of most substances can be well cut by soaking them in the india-rubber cement, which must be allowed to dry; the sections may then be made with a razor or scalpel.

For the observation of any object the student must place the same between a glass slip and a piece of thin glass. Water is the fluid most generally used for rough observation; but this must be left to observation and experiment. And in the mounting of objects common sense must be used; for instance, an opaque-looking object is generally best mounted in balsam, as it has good refractive powers, and a transparent substance is generally best seen when mounted either dry or in fluids.

Dust must be carefully kept from all preparations whilst in progress.

The author must now conclude, trusting that his readers will find these rough notes useful, and that the study of some of the hidden forms made by Divine art will lead him to search further for the marvellous beauties of nature.

Preparing the specimen for examination was only half the battle, of course. Once the slide was ready and the microscope to hand, there was always the problem of illumination. The developments in this ostensibly unrelated field had a continuous effect on the techniques of microscopy, and the gradual improvement in lamp design has been largely responsible for the self-contained microscope of today. Strange to say, most of the great discoveries of the past were made with the aid of oil lamps—their dim, yellow light makes that difficult to appreciate in the latter half of the twentieth century—and even electric lighting, when it did appear, took a long while to gain acceptance. Even in the 1920s it was exception, and not the rule.

To begin at the beginning, natural illumination is clearly the most easily available to the microscopist, and sunlight is the brightest and most readily obtainable in theory—even if it cannot be relied on for on-demand availability in northern latitudes.

Swammerdam used sunlight for all his observations on insects. It is related how he worked from dawn until noon, in spite of the heat, bathed in sweat and sunburned. His quest for knowledge was insatiable. Indeed, Swammerdam did not even wear a hat to temper the sun's rays. The broad brims cast an obscuring shadow and for more than a century afterwards it was regularly taught that one possible cause of dim lighting of the field of view was the obscuring effect of hat or wig.

Hooke took the microscope indoors and as we have seen he introduced the notion of artificial light concentrated by a large condenser lens on to the specimen. But this was only an extension of conventional viewing habits, for as we have seen, all of Hooke's drawings were of objects illuminated from above, ie by reflected (rather than transmitted) light. Leeuwenhoek pioneered the concept of conventional transmitted illumination, thereby paving the way for the method that was used in the bulk of important microscopical discoveries ever since.

It was Lieberkuhn who cast sunlight on to the specimen by means of a concave mirror in the manner originally propounded by Descartes in 1637. His detailed descriptions of the injected blood vessels in the intestinal villi were carried out using this method. Indeed the mirror device (which he introduced in 1738) soon became known as a 'Lieberkuhn', and it was a vital element of the so-called compass microscope which became very popular in the eighteenth century.

But lighting could not make much progress without advances in science. By the beginning of the nineteenth century, transmitted light was beginning to reappear as the technique of orthodox microscopy, and whereas the pioneers had tended to hold their simple lenses up to the sky, the microscopes that appeared in the early 1800s were fitted as a matter of course with a substage mirror. There had been many earlier attempts, indeed Huygens drew a rough sketch in 1679 which to all intents and purposes might as well be of a modern student's microscope—and the substage mirror is clearly in evidence.

The first attempts to provide an artificial source of illumination at a high level of brightness—something nearer to the sun's rays than a couple of candles or an oil-lamp—were the use of limelight in the early years of the nineteenth century. Lime was heated in an oxy-hydrogen flame to incandescence and provided a brilliant white light source. In 1845 the first arc-lamps appeared, though they were powered by arrays of cells that were costly and unreliable for many years.

In spite of the tendency towards technically advanced illuminants, conventional paraffin lamps remained popular for most purposes. The first improvement in design came in 1784, when a Swiss manufacturer, Aime Argand, produced the circular hollow wick that for many years bore his name. Air was thus admitted to the inside of the wick as well as the outside, and the positioning of a glass chimney around and above the flame caused a sufficiently energetic updraught of air to initiate almost total combustion of the volatile oil and the minimum production of smoke.

A Frenchman, M. Franchot, in 1837 improved the design by installing a piston activated by a spring which exerted pressure on the surface of the paraffin oil, thus forcing it up to the wick through a tube in the centre. Later in the century the flat wick still seen in many oil lamps appeared, and was retained by microscopists for many years. Duplex and triplex lamps were those with two or three wicks respectively, mainly intended as magic lantern projectors, but also used for high-powered microscopy. Gas lamps were popular too by the end of the nineteenth century, and were generally of two types: the batswing burner (still used for bench work in the laboratory) in which gas burned from a slit across the end of a metallic tube; and the fish-tail in which two jets of gas were forced against each other, spreading out into a flat sheet of flame. Of course, electric lighting began to appear too, but the deep-rooted acceptance of oil and gas lamps meant that electric filament lighting did not gather momentum for many years. At the beginning of the twentieth century, wax candles cost 34s per thousand hours; gas lighting gave around 8 candle-power and cost 8s for the same time; electric lamps gave about 16 candle-power and were said to cost about 40s for that period; and electric arc-lamps—1,500 candle power—averaged out at £15 per thousand hours. However, as the salesmen were quick to point out, that was by far the cheapest rate per candle!

Though the technology improved over the years, even in the 1920s the paraffin lamp was described as the microscopist's most useful light source. By then, the electric lamp had greatly increased its efficiency (it was at least ten times brighter than the earliest filament lamps used before the turn of the century) and of course the arrival on the scene of the incandescent gas mantle gave a new alternative—though the uneven nature of the illumination caused by focussing on the mantle fabric itself was a drawback.

A summary of the techniques then in use was published in 1921 by Conrad Beck in a short volume entitled, simply, *The Microscope*:

> For those who have not gas or electric light, but who require a more powerful light than a paraffin lamp, an extremely useful lamp, which is quite simple to use and gives excellent results, consists of an incandescent mantle heated by a methylated spirit flame. The reservoir having been filled with spirit, the method of lighting the lamp is as follows. The cap of the reservoir must

be screwed off, and the bellows attached by screwing in the nipple at the end of the tube. The bellows must be squeezed till the burner is hot. The U-shaped metal piece covered with asbestos should now be soaked in spirit and placed on the supporting tube below the burner, and ignited. This will heat the burner which is inside the hanging incandescent mantle. When the asbestos-covered U-piece has almost burnt out, the bellows should be gently squeezed two or three times, which will drive the spirit from the reservoir to the burner, where it will become volatilised and burn with a steady flame. The bellows may be greatly squeezed every five minutes if the light appears to be failing. The handle below the burner regulates the air supply, and should be adjusted till the best illumination is obtained.

The electric arc lamp is useful for photomicrography or projection, but is troublesome for general use.

The ordinary electric incandescent lamp provided with a frosted or ground glass bulb is a handy lamp for ordinary observations, but is not sufficiently brilliant for many purposes. It is supplied on an adjustable table stand.

The best equipment is the 'Pointolite', or 1/2-watt 'Grid' lamp, with a neutral glass double wedge, a set of colour screens, and a bull's eye. It does everything that is required for every class of illumination; and has adjustment so that the beam of light can be placed at any height between 3 and 9 inches above the level of the table.

If the condenser is not required it can be swung to one side; or if it is required to use colour screens alone, the lens of the condenser can be removed from its mount.

The lamp is provided with a ground glass and a signal-green glass; it is completed by the addition of the Wratten & Wainwright's colour filters and the neutral glass moderator. It is provided with 12 feet of cable and an attachment for fitting it to a lamp fitting of an ordinary house supply. For use with the 'Pointolite' lamp, which is an incandescent disc about the size of a small peppercorn, a direct current of any voltage from 100 to 250 volts is equally satisfactory, a variable resistance being supplied to adapt it to any current between these limits. The candle-power is 100, but as it is well concentrated in the one point it is at least twenty times as powerful as the filament lamp focussed with the condenser.

If a 100-candle-power 1/2-watt lamp or 40- or 60-candle-power metal filament lamp is used, it is suitable for either direct or alternating currents, and for a voltage from 100 to 200 volts, although a lamp suitable for the voltage must be selected. No special wiring is required, ordinary house current being sufficient.

An incandescent gas mantle, either of the ordinary or inverted type, makes a good light. Its only disadvantage is that it cannot be conveniently used with its image exactly in focus, because the fine mesh of the mantle does not then give a continuous surface. This light is sufficiently powerful for high-power dark-ground illumination if dark colour screens are not used.

A paraffin lamp with a flat flame is probably the most convenient light for general purposes, but it is not powerful enough for the use of colour screens or for high-power dark-ground illumination. When direct light is being used through a condenser, it is damaging to the eyes if too strong a light is employed. A strong illuminant is necessary for high-power dark-ground or opaque illumination, but it must be modified when direct light is thrown through the object. Some colour screens require a strong light, but immediately they are removed the light should be cut down. Just enough light to show the object readily should be used, and no more. If this precaution is taken, microscopists need have no fear of injuring their eyes, however long they work. The light should be more powerful than is required for general purposes, it should be powerful enough for dark-ground illumination and to allow of the use of colour filters.

A good form of paraffin lamp for microscopic work has a single flat wick 5/8 inch wide. The burner has a revolving motion and may be used with its edge facing the mirror to give a strong illumination, and with the flat surface facing the mirror for a softer light. It has a means of raising and lowering it from the table to enable it to be used for illuminating opaque objects with a bull's-eye condenser or parabolic reflector, or for setting it to the correct height for using a vertical illuminator.

The reservoir and burner are carried on a support which passes through the centre of the reservoir so that the weight is well balanced over the centre of the stand. The lamp glass is simply a 3 x 1-inch microscope slip carried in a thin metal chimney. The burner is insulated from the reservoir by a fibre ring, which is always cool enough to touch for turning the burner round. The metal chimney can be removed and the burner hinged back for trimming the wick. The reservoir has a large screw stopper for filling. A bull's-eye condenser on a separate stand may be used in combination with this lamp for illuminating opaque objects or for high-power dark-ground illumination, although this lamp is not recommended for the latter purpose.

Since that time, in the 1920s, the developments in technology have played a considerable part in making microscopy a more convenient and reliable occupation. Coiled-coil filament lamps took pride of place for many years, but after the 1939-45 war a range of new devices appeared. Discharge tubes of many kinds were introduced and the quartz-halogen lamp gave a bright source for many applications.

But, if the microscopist of fifty years ago would find it difficult to feel at home with the paraphernalia of sophisticated lamp units, he would notice very little difference when it came to the preparation of the microscope slide itself. Once the glass slide itself had appeared, staining techniques quickly developed and any microscopist who was working at the turn of the century would feel very much at ease with the routine techniques used for the preparation of specimens today.

In the days of the ivory slider, specimens were not pre-treated at all. And, apart from the injection of blood vessels which dates back to the earliest days of the subject (p24), the idea of colouring a specimen in order to accentuate certain structural elements was unknown. There was, early on, just one microscopist whose published work shows that he used a chemical dye in the preparation of sectional material. This was J. Hill, who carried out a systematic study of wood anatomy. His work, published in 1770 in London under the title *The Construction of Timber*, showed that he used carmine to tint specimens red. It was far ahead of its time, and there were apparently no followers along the path; but Hill remains the earliest known user of a stain, and, it is unlikely there were any precedents so early in the subject.

Over half a century later, C. G. Ehrenberg—a German protozoologist who worked at Leipzig—used coloured dye particles (carmine and natural indigo) to trace the movements of food vacuoles in ciliate organisms. This method had been up-dated and is used today. Indeed the addition of pH-sensitive materials which change colour with an alteration in hydrogen ion concentration, has enabled us to learn something of the 'digestive cycle' in protozoa.

Ehrenberg's work was published in 1838, and carmine next

appeared in the literature in 1849. Two botanists, Göppert and Cohen, used the dye to colour cells of *Nitella* that they were studying. There are interesting cyclical movements in the cytoplasm of this plant which are easy to study as the cells themselves are large. Two years later the histologist Corti (after whom the organ of Corti in the inner ear is named) used carmine in a 50 per cent alcoholic solution to stain (and, though he did not realise it, also to fix) squamous epithelial cells from the cheek mucosa. Another botanist, Hartig, took the process a stage further by using Corti's stain for leaf tissues taken from green plants, where he found it of value in staining the plastids. This was around 1855—and at that time two important steps were dramatically taken.

The first was the work of J. Gerlach, a German histologist. He found that carmine mixed with ammonia (which forms ammonium carminate) gave all the signs of being a very much better stain than carmine alone. He set out to test the idea and found that strong solutions of the material would certainly dye the specimen—but the colour was so strong, if it took at all, that no additional structures could be perceived. Weaker solutions, on the other hand, did not stain the tissues successfully. It was then that fate took a hand. One night, tired after a round of unsuccessful experiments in the laboratory, Gerlach left his reagents where they were overnight, instead of tidying up. Amongst the materials left on the bench was a jar of dilute ammoniated carmine—and standing in it was a section of nerve cord. When he came to clear away the following day, Gerlach noticed that this section—which had been previously fixed in bichromate—was clearly stained. The histological configuration was clearly revealed in a well-differentiated manner. And this was the first 'stained histological section' (using the term as we would understand it today) in microscopical history.

But, as Gerlach was working in Germany, another important series of experiments was under way in England. They had nothing to do with microscopy at all, as it happens, though the eventual marriage of the two arts was to prove vital to the progress of science. For this was the moment when artificial dye-stuffs were to emerge—and they did so because of the genius of one man, W. H. Perkin.

William Perkin was born on 12 March 1838, the son of a building contractor. He later studied chemistry at London, but worked mainly in what he described as 'my rough laboratory at home'. Stimulated by the experiments of his lecturer, Perkin decided to try some experiments involving the oxidation and hydrolysis of aromatic molecules. His first experiments at home centred on aniline sulphate, which he treated with potassium bichromate. The result was a black precipitate. Perkin dried it and decided to look for a solvent for the residue. He found it in alcohol, which readily dissolved the black grains to form a brilliant purple solution. During the rest of the Easter vacation, Perkin tried to use the material as a dye for pieces of silk. He found the colour (due to its lack of solubility in water) to be fast and, what is more, it resisted fading from sunlight. Working with his brother during the summer term, he eventually applied for a patent on the idea and then, against the advice of his tutors, he left college—without staying to qualify—and set up his own manufacturing laboratory. He was still only eighteen...

This was the very start of the aniline dye industry, and within a few years many manufacturers in Germany and France had taken up the idea for themselves. The word 'mauve', now a normal part of English, was in fact originally a trade name used in France for Perkin's purple dye, when manufacturers—realising there was no valid patent protection of the process in France—began to produce it in quantity. By 1859 production in Perkin's factory was on a commercial level; ten years later his annual output totalled a ton of the dye. But then the continental interest mushroomed dramatically; in 1871 a few dozen kilograms of the dye was produced by Gräbe and Liebermann in Germany—but just three years later their annual production had reached a total of over 1,000 tons. Perkin's yearly total was not even half this.

But meanwhile the availability of these textile dyes was noticed by microscopists. In 1858 fuchsin was discovered in France, and within three or four years it was first used as a microscopical stain. By 1860 red, violet and purple aniline dyes had all been tried, with varying degrees of success, and then in 1865 Bohmer introduced haematoxylin as a histological stain. This material has, of course, remained a favourite cytological stain for routine work ever since.

Strictly speaking, haematoxylin is *not* a dye at all; it is quite colourless. This material is extracted from the heart wood of *Haematoxylon campechianum*, a central American leguminous tree, and it is readily oxidised to haematin—which is the stain used in microscopy. The haematoxylin that is purchased for use is a brownish residue containing a mixture of partly oxidised haematoxylin, and most recipes for the staining solutions (used today) make provision for the oxidation of the material by atmospheric oxygen. But it is worth noting that the widely-used haematoxylin staining solutions are nothing of the sort; if they were they would stain nothing.

The most popular stains used in Britain by 1870 were known, after the manufacturer, as Judson's Colours. They were very popular, but suffered from a serious drawback in that they tended to dye everything indiscriminately. In 1869 the concept of differentiation (ie the removal of surplus stain by solution in alcohol) was first raised in the literature and by the mid 1870s the idea was becoming widespread. In the 1880s several manufacturers—notably Dr G. Grubler of Leipzig—began to manufacture smaller quantities of aniline dyes of greater purity and standardisation, solely for the use of histologists; and by the turn of the century the range of stains (with the exception of the fluorochromes used in fluorescence microscopy) was extensive. The permanent preparation of today is generally identical with those dating from the turn of the century.

CHAPTER FOUR

# THE 'MODERN MICROSCOPE' ARRIVES

The conventional microscope had reached its peak in the late nineteenth century. Since then it has settled down to a position somewhat less sophisticated in terms of accuracy and finish, and certainly not as precise in mechanical terms as some of the Victorian instruments; but it is a reliable and relatively foolproof piece of equipment. The development of the electron microscope has allowed high-resolution work to go on in a new discipline of its own, and so the pressures on the modern microscope manufacturer are for ease of operation and consistency, rather than ultimate resolution or technical intricacy.

## THE RESEARCH MICROSCOPE

The typical microscope used in modern research has acquired several features that were rare at the turn of the century, and it has discarded several others. A binocular eyepiece assembly is usual. This is not a new idea—on the contrary, Cherubin d'Orleans described and drew a binocular instrument in the 1670s, and binocular microscopes were built in the late 1800s after a good deal of research in the United States. Louis Pasteur is reputed to have used a binocular microscope when he worked at the Whitbread brewery in London around 1870.

But it is only in the past decade or two that the binocular microscope has truly come into its own, and now virtually every

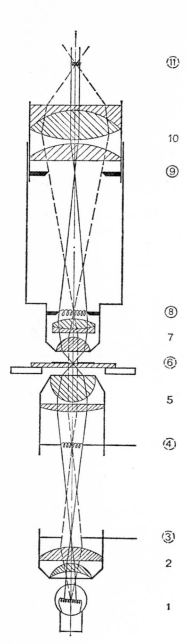

Fig 6 Passage of rays through a compound microscope (not drawn to scale). 1 minimum size of light source; 2 lamp condenser; 3 diaphragm on lamp housing; 4 condenser iris diaphragm; 5 substage condenser; 6 specimen plan; 7 objective system; 8 level of iris diaphram image; 9 eyepiece (=field) diaphragm; 10 compensating eyepiece; 11 position of observer's eye and plane of lamp image.

researcher worth his salt has one. Every routine microscope now has a coarse and fine adjustment for focussing, but few include in the design a centering substage for normal optical work. Fine adjustments on the condenser are never normally found, and nor are many of the fitments that characterised microscopy until the 1930s—live-boxes, compressoria of one kind or another, stage forceps and so forth. Many microscopes now have a light source built into the base somewhere, thus obviating the need for a mirror and an external lamp, and many models on sale do not even have the facility for fitting a mirror, even if one was required. Nor do modern microscopes feature an adjustable tube-length as a rule. It is tacitly understood that microscopists today do not appreciate the niceties of carefully controlled correction, but prefer to use the instrument as it comes, and utilise regular coverslips for all purposes. So the dropping of tube-length calibration is a pragmatic trend—a reflection of the times.

There has been a tendency towards the incorporation of concentric controls; mechanical stages and focussing controls are often mounted on the same shaft. This principle originated with the mechanical stage designed by Turrell in 1833, but only in recent years has it come to predominate in the design of microscope controls as a whole. The typical focussing adjustment has a large milled head for the coarse adjustment, and a smaller one for the fine focussing mounted against it. The iris diaphragm has tended to lose favour, too, as rheostat-controlled lamps are now so widely used. But in many designs the moveable body-tube has been replaced by a vertically sliding stage which is operated by the focussing controls. In this way focussing is carried out by raising or lowering the stage, whilst the optical assembly in the body-tube remains firmly anchored to the main stand. Finally, an inclined eyepiece assembly has become almost obligatory, since this eliminates the need for vertical observation (which is tiring) or, alternatively, tilting back the whole instrument (which can cause a liquid preparation to 'run'). Not only is observation facilitated by this development, but the slide is *always* horizontal

—and a vertical tube can be installed on which a camera can be mounted, above the observer's forehead, quite out of harm's way.

And as the optical arrangements have been modified with ease of operation in mind, microscope stands have tended to turn round, too. The limb in the traditional instrument was nearest the microscopist; now it frequently faces away from him. This puts the focussing controls (mounted near the base of the stand) a comfortable forearm's length away and makes for a restful stance. However, since 1965 there has been a further refinement (initiated by the design of the Vickers' 'Patholux' stand) in which the traditional limb is replaced in entirety by a heavy rectangular alloy casting. This makes the microscope look more like a solid, upright cabinet with stage and optical equipment 'hung' on the front. The controls are situated at the base of the heavy foot, nearest to the observer.

Microscopes arrive with a printed guide detailing each model's design features, which should be kept available for consultation; and this book would quickly deteriorate into nothing more than a catalogue of maker's information if we tried to do justice to such details here. However the description above should enable a microscopist to find his way around the typical research microscope of today.

## 'MINIATURISATION' IN DESIGN

Not all modern instruments are 'typical research microscopes', however, and it is important to consider the exceptional designs as well as those more orthodox if we are to obtain a balanced concept of contemporary microscopy. One difficulty with any design has been size. Microscopes are, at best, heavy; and at worst they are impossible to transport from one site to another without a great deal of packing away before the move, and adjustment after it. There has been a need for a truly portable microscope—not merely a collapsible one, but a totally redesigned, truly pocketable instrument.

It was in 1934 that a British microscopist, J. McArthur,

announced his own revolutionary design for a pocket microscope. It was constructed as a die-cast metallic box roughly 4in long and 2in tall. Light entered through the top left-hand corner, passed down through the inverted slide, thence through the objective before being reflected by two prisms, only to emerge through an eyepiece at the top right-hand corner. A coarse adjustment was eliminated altogether, since the inverted slide meant that the upper surface of the slide (ie the surface bearing the specimen) always came to rest on a pair of solid supports. Thus, only differences in coverslip thickness of the amount of mountant could throw the image out of focus, and that called for only the slightest adjustment. The objectives were mounted in a sliding holder, and were carefully selected for parfocality; and experiments showed that the finished article could be run over by a car or dropped from an aircraft without damage.

But the McArthur microscope had to wait some 30 years before it became widely known. The founding of the Open University led to a demand for a supply of cheap, foolproof microscopes and the McArthur model fitted the bill admirably. A range of accessories for micrography and including built-in illumination attachments has been designed, and the McArthur microscope is now a unique part of the microscopist's armoury—unique because it is the first totally rethought design in the recent history of microscopy.

And, as that microscope is the smallest and most compact, we may by contrast look at the Burch reflecting microscope. Far from being small enough to slip into a pocket, this rare device takes two men to lift it...

## *THE LARGEST OPTICAL MICROSCOPES*

Reflecting microscopes are the cinderellas of the subject. There has long been a recognition of the fact that mirrors can be given a concave contour that magnifies an image—the modern shaving mirror, which is such a magnifier, has many antecedents —and the first practical design for such an instrument (almost an inverted 'short-focus telescope', if that is not a contradiction) appeared in 1736, and another design dating from a few years later looked as though it would provide the jumping-off point for a rash of similar designs. It fell to the lot of Gianbattista Amici, the Italian microscopist, to produce the first truly practical design in the 1810s. This was part of the move towards achromatism, but the later development of achromatic refracting objectives rather took the wind out of its sails.

It was in 1936 that a 34 year-old microscopist, C. R. Burch, returned to the subject. He looked into the work of an astronomical optician named Schwarzschild, who had worked at Gottingen at the turn of the century. The formulae that Schwarzschild had derived were intended for the *Spiegelteleskop*, the reflecting telescope, but Burch realised that they could as easily be applied to microscope design. Thus he drew up specifications for mirrors with an aspheric contour (ellipsoidal, near enough) which would give an image free of any chromatism and also, of course, with excellent correction of distortion. He constructed the mirrors out of speculum metal. They were ground, then lapped spherical, after which the peripheral regions of the mirror were formed with oversized laps to give a rough approximation to the Schwarzschild calculated curve. The mirror (still marked with concentric 'steps' because of the lapping procedure) was then placed in the well of a specially designed figuring machine which by lightly polishing the surface with rouge brought it to its mirror-smooth profile.

Periodic tests were made on the progress of the polishing, using an interferometric method which readily demonstrated any surface irregularities. The final profile was correct to within a fraction of an optical wavelength, and the whole figuring process took a very considerable time.

The Burch mirrors were mounted in a heavy microscope body manufactured out of brass, steel and alloy; the stand alone weighed over a hundredweight. Ten of these microscopes were made by the Clevedon company of Willcocks Ltd. They were expensive, but they have since become part of the folk-lore of microscopy. One of the best-known of these instruments was commissioned by the Chester Beatty Research Institute and used by Dr E. M. Roe in a lengthy programme of research into cancer. Dr Roe tragically died (of cancer) in 1971 in the Royal Free Hospital, to which her Institute was attached, and it was with considerable excitement that I learned of the Institute's decision, at her suggestion, to donate the microscope to my own laboratory where we were pressing ahead some work in the field of blood coagulation, where the use of ultra-violet light was of particular value. The Burch, of course, can be used in the U/V region of the spectrum as well as in the visible region—and once in focus for one colour it is equally in focus for any other wavelength.

The pressures of developing a unique instrument of this sort weighed heavily on Burch's shoulders. The building of the aspherizing machine itself, for the production of the mirrors, was a tedious and delicate process. The effects on the microscopist were profound. Burch took to drugs. The teaching duties he bore during the day, when he was with his PhD students, prevented him from concentrating fully on the task in hand, and the stimulants kept him awake as well as reviving some of the details of the previous night's work. After five years Burch was forced to leave the project altogether. It had become abundantly clear, he wrote to me later, that he would die if he didn't. Some years later, indeed, he did become seriously ill with a perforated gastric ulcer, and it took another four years for him to recover and, at the same time, to start to understand the pressures on his intellect that had lead to the collapse of personality he had suffered.

Such are the strains that can be placed on the innovator. Burch was not unique; others have found the pressures to be considerable too. And it is important for us to realise the effects that microscopical research *can* have; it is deceptively easy to imagine that this is a gentle, methodological field in which to work. But even microscopy, it seems, can exert a weight on the personality that it is customary to associate with high-powered executives at the pinnacle of a large company in commerce...

Though the modern optical microscope ranges from the massive precision of the Burch down to the inexpensive, pocket-weight McArthur portable, the run-of-the-mill routine microscope remains essentially based on the concept outlined on p77.

## PHASE CONTRAST

We have to face the fact that the conventional microscope leaves much to be desired. In one field in particular, namely the examination of living material, it has manifest drawbacks. Living matter is, as a rule, more or less transparent. Its essential constituents, cytoplasm and the rest, are dispersed in water and they have a refractive index only marginally different from that of the aqueous mountant by which the living cell is surrounded. As a result the light rays transmitted by the specimen are not materially refracted from their course, and only the faintest of images is formed.

However there have been other effects on the light beam that are not visible in the ordinary way. The diffracted wavefront may have been retarded in phase, even if it has not been significantly refracted. It is possible to demonstrate this apparently slight difference by combining the transmitted ray with the unaltered reference beam. Light that has been retarded sufficiently to render it exactly out of phase with the reference beam will effectively 'cancel out' the amplitude of the resultant ray, and so the object will appear virtually black. Yet this startling change is due to a retardation of the illuminant by only one-half wavelength!

The visual effects of phase contrast are very useful: even a slight alteration in optical density of the specimen causes a significant alteration in the brightness of the observed image, and small (virtually invisible) structures can be clearly discerned.

The method was originally mooted by F. Zernicke in 1935. He had been engaged on studies of telescope mirrors and found that phase changes could be used to test them for surface configura-

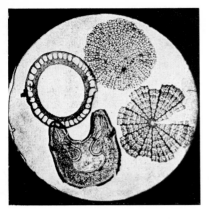

Page 85 (*top*) Four plant stem sections (one a petiole) photographed by Fox Talbot *circa* 1841. They are the first examples of true photomicrography. Note the traces of Canada balsam mountant at the edge of the circular field of view, although the specimens are dry mounted

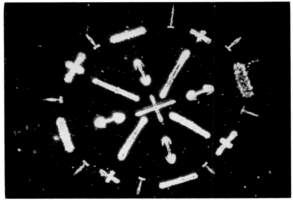

(*centre*) A popular Victorian pastime was the mounting of siliceous spicules of sponges in intricate patterns. This example is mounted against a black surface and is seen by incident lighting

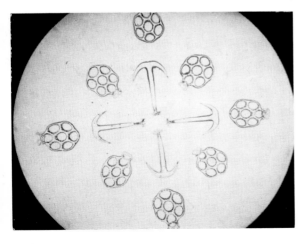

(*bottom*) Another example of spicule mounting. This specimen, mounted in balsam, is illuminated by light-ground

Page 86 (*top*)  Microscope slides from 1720-1920. Note ivory sliders (top left) and small mount of flea between two glass sliders next to them. Several paper-covered glass slides from the period 1860 are shown. Note also the cardboard slide (lower left), sometimes used for opaque specimen mounts. Serial sections on a larger 3in x 1½in slide are shown, below right. (*bottom*) Two centuries of technical improvement—early eighteenth century appearance of squamous epithelial cell (left) viewed with Wilson screw-barrel microscope, compared with early twentieth century view (right). Note that the largest bacterial cells were just resolved by the Wilson lens

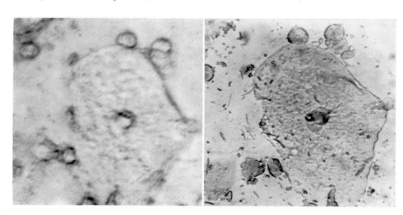

tion. In the following year he extended the principle to microscope work, and realised that the phase changes brought about by the object might be used as a means of image contrast-enhancement. There had been workers before him who had carried out phase observations. But they had been concerned with the anomalous effects that unusual lighting conditions could bring about, and there was no direct precedent for Zernicke's interesting new idea.

This was to retard the zero-order beam (ie the direct beam that does not pass through the specimen) by $\frac{1}{4}$ wavelength ($\frac{1}{4}\lambda$) so that slight changes would bring about either a brightening or a darkening of the image structure. At first he retarded the zero-order beam by means of a phase strip—a piece of glass of a thickness calculated to bring about the desired phase change. Zernicke took out patents on the process in collaboration with the German optical manufacturers Zeiss, who produced a commercial instrument in 1941. But work did not press ahead until after the war's end. In 1942 a paper on the method (written, incidentally, by C. R. Burch) appeared in Britain, and shortly afterwards the first successful films of living cells were produced by using the technique. An upsurge of interest followed, and now the phase-contrast microscope is a standard part of the microscopist's armoury.

Clearly the phase 'settings' are arbitrary. Zernicke proposed two definitive standards:
(a) equipment where the zero-order beam is advanced in phase relative to the transmitted light—POSITIVE PHASE
(b) equipment in which the zero-order beam is retarded in respect of the transmitted light—NEGATIVE PHASE.

Though this terminology is now generally accepted, it is noteworthy that exactly the opposite convention has been used in the United States by some microscopists. It is also important to realise that the phase plates themselves do not determine the appearance of the specimen which may be either: when the object is darker than the background—DARK CONTRAST, when the object is brighter than the background—BRIGHT CONTRAST. This depends on the relative refractive indices of the specimen and its mountant (usually a saline solution for living cells, or water for micro-organisms). If the relationships are interchanged, by the choice of a different mountant, then the contrast effects will be reversed also.

Yet the system still embodies several important drawbacks. In the first place the image is seen to be surrounded by a prominent line of high contrast, an indistinct 'halo' that obstructs the clear study of certain particulate structures. Secondly, there is a concomitant loss of resolution.

And, thirdly, though the inevitable phase changes incurred in biological microscopy are usefully exploited in the technique, there are no means of controlling the phase retardation, and therefore no accurate way of quantifying the results.

## THE INTERFERENCE MICROSCOPE

To meet the above requirements, the interference microscope has begun to gain popularity in the past few years. Since there is some confusion in many minds over the distinction between this form of instrument and the phase-contrast microscope, it is as well to state at the outset that both are interference microscopes; in both the image is formed as a result of interference between a reference beam and a transmitted beam. But whereas the beams are both transmitted by the object in phase microscopy, the image contrast resulting from interference between the zero-order and the diffracted rays, in the true interference microscope, is owing to interference between light retarded to varying degrees by the specimen, and a quite separate beam that has by-passed the specimen altogether. The phase of this latter beam, by the use of an optical glass wedge, can be retarded by any desired amount, so that the contrast and hence the optical properties of the image's substructures may be studied. In this way structurally related features may be identified, and phase shift readings may be taken to give an index of refractivity. Not only this, but the annoying fringes seen in the earlier phase-contrast system are avoided and the resolution remains high.

The reference beam is kept apart from the specimen by one of two methods. Either it is brought to focus after the transmitted beam (ie above it)—a form known as the 'double focus' system;

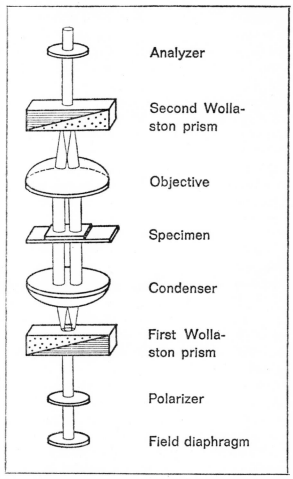

Fig 7 Idealised diagram of a split-beam interference microscope

or it is brought to focus alongside the specimen but in its plane, the so-called 'shearing' system. Both are available commercially. Alternative designs use either a separate by-pass microscope—the beam being split, and half becoming the reference beam whilst the other half passes through the specimen before recombination —or a Dyson interferometer which uses lightly aluminised mirrors to divide the illuminating beam and recombine it in a device that looks like a larger-than-normal objective housing. The double microscope system is suitable for large specimens and obviously can be calibrated for very precise quantitative work, whereas the single systems necessitate a smaller specimen area (or the reference beam would be interrupted), but are simpler to use.

Other interference microscope systems have been devised, such as Pluta's experimental 'amplitude contrast' apparatus, and the Nomarski interference contrast optics produced by Zeiss. The latter gives a dramatic illusion of three-dimensional 'shading' in biological specimens which can be used to aid structural interpretation.

### LASERS AND HOLOMICROGRAPHY

The interference patterns set up by laser-illuminated specimens provide the principles of holomicrography, in which the hologram, when illuminated by the laser after processing, enables the specimen to be visualised. Experimental work indicates that a range of illumination techniques—including light-and-darkground, and phase—may be simulated by careful control of the laser visualisation. A further refinement is to obtain the holograph as a 'contact print' of the specimen, subsequently magnifying it optically. But these are techniques for the future; and they remain of only academic interest to the practising microscopist.

### DARK-GROUND MICROSCOPY

Another approach to contrast enhancement in the conventional microscope is to observe only light reflected by the object, so that

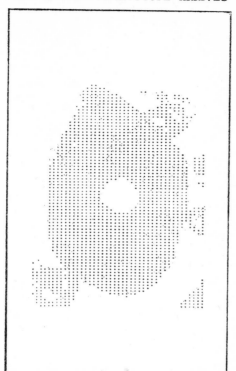

Fig 8a  High-speed scanning of an epithelial cell shows size and shape of nucleus. Similar cells seen by light microscopy are shown on p86

the field of view itself is otherwise dark. This is known as darkground microscopy. The simplest means of obtaining the effect at low powers is to use either strongly oblique illumination, or to insert a stop in the substage in place of the condenser, in such a way as to screen out only the peripheral rays. The direct—central—beam is blocked by the patch-stop, as it is called, and the image may be seen to adopt a silvery, luminous appearance against a black background.

The technique is of particular application to the visualisation of small structures, which may reflect enough light to be visible

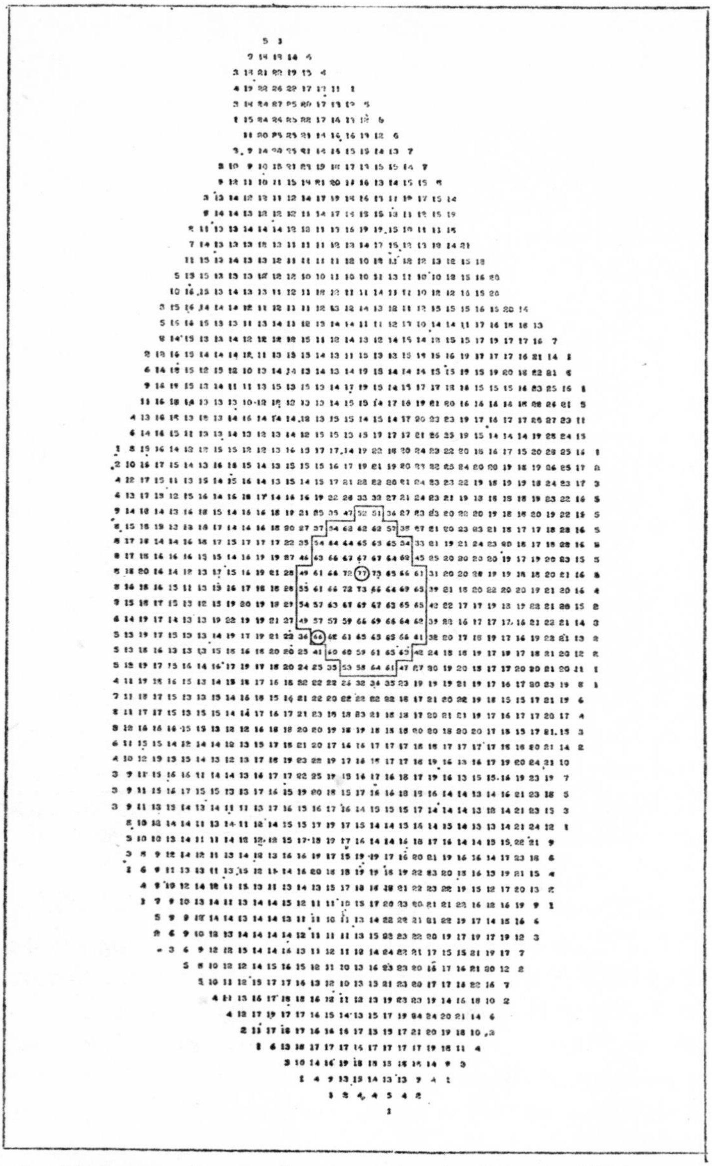

Fig 8b  Computer print-out of percentage transmission in different regions of the epithelial cell in Fig 8a. Coded in this numerical form, the visual appearance of the stained cell is amenable to statistical analysis (see also p86)

even if they are too small effectively to set up enough diffraction disturbance to register as a dark image by conventional light-ground illumination. In this form it is known as an 'ultra-microscope'. I have used it in studies, for instance, of the fine structure of a blood clot; it enables one to visualise fibrils that are too fine to be resolved in the ordinary way. The term 'visualise' is used here because it removes the ambiguities that arise if we talk of 'resolution'. Since proposing this term some years ago I have been gratified to see it gain acceptance, as it does help to remove ambiguity. The mathematics of the visualisation mechanism would take us beyond the scope of the present discussion—but it is important to realise that these very fine structures that are beyond the resolution limits are of a size that depends on the aperture of the lens (which cannot be more than about N.A.=1.00 for dark-ground work), and not on the dimensions of the fibril. The diameter of the structures varies with their apparent brightness or reflectivity—so measuring the dimensions of micrographic records gives no information about the diameter of the structures. It is an important fact to bear in mind.

Dark-ground studies are often highly aesthetic. The pictures of erythrocytes suspended by their microfibrils in the human coagulum—penderocytes, as they are called—which I took several years ago, show the clarity of the results that can be obtained. This micrograph is of the highest magnification that has ever been obtained with the optical microscope—yet it is surprisingly clear and detailed.

Another form of dark-ground microscope results from the use of crossed polarising screens, one below the condenser, the other in the eyepiece. If rotated to give extinction of the background, mineral or other optically active materials will appear rainbow-tinted against the dark ground. This is a technique widely used in crystallographical microscopy.

## *THE INVISIBLE WAVELENGTHS*

But the optical microscope is not only used with visible light.

The electromagnetic spectrum can be used when it extends beyond the wavelengths that are detected by the human eye. Ultraviolet microscopy has shown details that are finer than those that can be resolved by visible wavelengths, and infra-red microscopy—though a relatively unheard-of technique—has important applications as well. Infra-red, since it is greater in wavelength than visible light, clearly cannot improve resolution. Rather, it diminishes it. But for low magnification microscopy this is a factor of no importance, and the special characteristics of infra-red can be used to advantage.

Many coloured materials—such as insect chitin, commercial dyes, etc—absorb strongly in one or other of the visible wavelengths but not in the infra-red. This means that strongly-coloured objects may be examined as though unstained; and the lack of contrast with the corollary of increased visible detail is sometimes useful. Whole insect mounts may be photographed in this way with the revelation of much internal detail. On the other hand, some materials (such as blood) show strong absorption of infra-red, and the use of this illuminant serves to increase contrast in an otherwise faint specimen.

It is important, in using infra-red sensitive materials, to realise that these wavelengths can penetrate some soft materials in which cut films, etc, may be stored. Film can be accidentally fogged in this way, and tests are advisable before sensitive materials are kept in such containers.

Ultra-violet wavelengths were formerly widely used for increased resolution, but the use of the electron microscope has made this unnecessary in most cases. Probably the most practical method of micrography using ultra-violet was the Bausch and Lomb system, which is now no longer in production. A range of objectives was produced which were achromatic and had a common focus for wavelengths of 0.546 $\mu m$ and 0.365 $\mu m$—the green and ultra-violet lines of the mercury arc. This meant that the image could be focussed visually, and then the green could be eliminated by the insertion of a U/V filter for the photographic exposure. Fortunately enough, the 0.365µm wavelength is transmitted by glass, so that no special quartz optical elements are required. Of course, any objective can easily be employed if the calibration difference between the two focussing positions is first ascertained empirically. Alternatively a fluorescent screen may be used for focussing.

Greater resolution can be obtained by using light in the far ultra-violet, for which specially made optical elements are required and—if very small wavelengths are to be used—the whole microscope has to be placed in a vacuum chamber. The use of a reflecting microscope ($qv$) removes the difficulties of focussing, of course. But these are aspects of a very specialised nature that have only limited applications.

## *FLUORESCENCE MICROSCOPY*

The most widespread routine and research use of ultra-violet is in fluorescence microscopy. The principle is, simply, that certain chemical materials will absorb light of a short wavelength and emit it at a greater wavelength—ultra-violet may be emitted as orange, or green, for instance. Such a material may be bound to specific parts of biological structures and the site of the labelled material can then be revealed by the use of fluorescence microscopy.

The first attempts to utilise this phenomenon date from the first few years of this century, when A. Köhler noted primary fluorescence (ie luminescence of the specimen itself) during his experiments with ultra-violet microscopy. In the 1930s many workers realised that fluorescent staining could show details of structure which—shining against a dark background—would easily be invisible by normal methods. Around 1940 the idea first appeared for the 'labelling' of antibodies with fluorescent dyes (known as fluorochromes) as an aid to location under the microscope. But it is only in the past twenty years or so that the idea has become important as an aid to research.

Primary fluorescence is used as an aid to the identification of certain chemical constituents in dusts or sections. It may be used to show a distinction between natural and synthetic minerals, for example, or to identify fragmentary material in forensic work. Secondary fluorescence is now used widely in antibody studies. Acridine orange, introduced in 1948, became the most widely used fluorochrome after it was found to distinguish between living and dead material *in vitro*. It was later found that it selectively activated DNA in the nucleus green, and cytoplasmic RNA red at around pH 6.0—it has been somewhat superseded by other materials including euchrysine.

In the use of a fluorescence microscope, it is important to realise several points:

The substage must be of material that can transmit U/V of the desired wavelength, but—since the excitation results in visible wavelengths—glass optics can be used for magnification.

U/V rays will be liable to enter the microscope direct unless a filter is present to prevent this: this can cause fluorescence of lenses or lens cements, or even severe damage to the eyes.

Blue light will, in many cases, produce an excitation of a similar nature and can usually be substituted for routine use with advantages in both ease of manipulation and safety.

## THE STATE OF THE ART

In this way the microscope has gradually settled down into its position as a tool of modern research. We must realise that it has not, in all respects, continued to develop as before: many features of earlier microscopes are now no longer available, and in some cases this leaves us in a position inferior to that of the microscopist many decades ago. No one misses the stage forceps that gradually died out between the first and second world wars, but the lack of attention paid to dark-ground apparatus, the poor quality of many objective systems, and the lack of operational flexibility in many modern microscopes is to be deplored. The use of ultra-violet light—with glass objective and eyepiece lenses —should be far more widespread than it is; and corrected objectives can eliminate any difficulty in focussing. Wider use of interference systems could avoid the serious drawbacks inherent in phase contrast, and there is no reason why reflecting optics should not be mass-produced.

So we are in an odd position. The most impeccable mechanical stands were produced a century or more ago; the best optics were available at the turn of the century; and the modern microscope —though a triumph of design for the manufacturer—often leaves much to be desired by the user. Little wonder that so many micrographs that appear in the literature are of an inferior quality. The teaching of microscopy is a very neglected art.

This state of affairs is anomalous in pragmatic terms too; for the most important discoveries in biology and the function of the human organism are yet to be made—and the microscopist will play a key role in making them. So perhaps the true pinnacle of the evolution of microscopy is yet to come...

CHAPTER FIVE

# THE MICROSCOPE IN ACTION

### *SETTING UP THE INSTRUMENT*

*The simple microscope*
Single-lens microscopes are not usually used these days. The most common form is, of course, the hand lens or 'magnifying glass'. In use, the best means of gaining a satisfactory image is to set up the object to be examined in diffuse (ie omnidirectional) light. Out-of-sun daylight is best, though fluorescent lighting or the light of a pearl electric lamp bulb are excellent too. The eye adapts to evening conditions, and artificial light is more than adequate. Indeed, as my experiments with early microscopes have shown, even a single paraffin-wax candle was a very good source of illumination for many purposes!

The lens should be held close to the eye, but not (as is often instinctively done) at arm's length. Then, the lens/object distance should be increased until the object is just about to go out of focus, a process that is quite instinctive in practice. The maximum magnification and clarity of image are in this way obtained.

Objects to be examined with the hand-lens by transmitted light are best viewed against white paper. The paper can be illuminated by direct sunlight. But it is important that the object (which may be a transparency, or a mounted microscope slide, for instance) is not in contact with the paper surface. It should be a few centimetres above it, supported on a pair of matchboxes or pencils, or even held in the hand. This avoids interference from shadows cast by the object on the background itself, which can confuse the interpretation of the image.

In using an early simple microscope, such as an example of Wilson's manufacture, the procedure is similar. The object slider is inserted and aligned with an easily identifiable specimen in position. A plant section, such as *Juncus* or *Sambucus* pith, is best. Whole specimens—such as a louse, flea or other small insect—are smaller than sections of this kind, and are correspondingly easier to miss even when they seem to be in front of the lens—particularly at high magnifications.

If a lateral source of illumination (such as sunlight) is to be used, a reflector such as a piece of white card, curved to act as a crude focussing surface, should be placed on the opposite side of the object in order to reflect light on to the shaded side of the specimen. This is particularly important for photomicrography, since the sensitivity of film to a wide range of degrees of contrast is factorially below that of the eye. A side-lit object which forms an acceptable image for the microscopist to examine direct, may make a print that is grey and lifeless, or which shows scarcely more than a barren silhouette. Some exceptions are discussed below, however.

If the microscope has built-in vertical illumination—for metallurgical work etc—it is a fairly easy process to switch this on, focus the image by racking up with coarse and then finishing with the fine adjustment, and check for alignment and brightness. The light level should be adjusted to give a comfortable appearance. But light on the bright side is preferable to that which is a little too dim. In metallurgy, since a metal specimen is brightly polished as a rule, and reflects light directly back into the objective, it is important that the specimen surface is exactly normal to the optical axis of the microscope and that the brightness and orientation of the illuminating beam are both carefully regulated. This process is aided (and considerably simplified) by the mirror-like nature of the object. When the illumination is correct, the object is brightly lit and the effect of maladjustment is clear to see—and, therefore, to eliminate.

With external illumination sources, problems arise as we move up the magnification scale. The light is spread out over an area that increases more or less as the square of the linear magnification (actually—where m=magnification—by $[m+1]^2$) and hence at even medium powers the image may become very dark and indistinct. Lamps of increasing intensity are used to offset this setback, and flash may be used when photography is the aim.

The selected lens is then inserted and screwed right home. If a compass microscope is used—or any other kind of simple instrument—the object/lens distance ($d_o$) is made very small. The two are brought close together, almost touching. As in the use of a hand lens, the purpose has been to have the lens *too close* to the object.

The lens is then unscrewed, or in whatever other way is appropriate gradually brought further away from the object. As $d_o$ increases, the image begins to focus. Final adjustments are made only when the object begins to go out of focus—when $d_o$ is now too great—until a clear image is obtained. It is important as a matter of course to go a little too far in making the initial focussing adjustment, so that the correct position can be obtained by back-and-forth movements; an attempt to creep up to the correct position of focus and stay there is doomed to failure from the start. Indeed a successful microscopist will have one hand on the focus control as he scans across a specimen, making continuous small adjustments in and out, so that the image remains clear and does not slowly drift from the optimum.

*The compound microscope*

The principle of focussing is the same in the use of the later compound microscope—the objective is brought within the focal distance and then racked up until the image is clear. But the first consideration is the question of illumination. It is not enough to use a pearl lamp and adjust the mirror, or to switch on a substage illuminator and leave it at that. Illumination is important; on it depends the clarity of the image, and too little attention is normally paid to this aspect of microscopy.

*Incident illumination* is applicable to the lower magnifications only. The image is viewed by *reflected light*, exactly as has been described above. Because the compound microscope gives higher magnifications than a hand-lens, artificial light is often used, in which case it is important to have the lamp shining down on the object as far as possible.

For special purposes, the general form of lighting described above may not be enough. Some specimens, for instance, require to be lit strongly from one side. When this is the case a small lamp unit may be situated alongside the object and raised or lowered until the correct level of shadow-formation is achieved. This form of lighting control is necessary when it is desired to show up the relief of a surface such as paper, cloth etc. The lower the angle of the light relative to the plane of the object, the greater will be the length of the shadows cast. It may be necessary to use a black drape opposite the lamp in order to absorb light that might be reflected into the shadows if a very hard relief effect is to be obtained.

On the other hand, very even illumination—intended to eliminate any relief effect, to reveal only surface tones and colours—may be demanded in applications such as print examination or micropalaeontology. Here a mirror may be used as reflector, to help balance the light distribution, but the most effective device of all is a ring illuminator. It is possible to purchase such a thing, though they are not very widely used and some workers prefer to make their own by mounting low-voltage lamp holders inside a ring of metal or plastic, several inches in diameter. This has the advantage of flexibility; a few lamps on one side can be unscrewed or screened in order to give a slight shadow if this is necessary for the observations in hand. A ring flash unit is ideal for micrography, and can easily be mounted for the purpose. And finally there are objectives available in which there is a concentric external reflector that directs light on to the object even at very short working distances.

For certain applications, such as the microcopy of textiles, it is helpful to use both reflected light and transmitted illumination. Here the specimen is lit from above as we have described, but it may be supported on a transparent or translucent screen which itself is lit from below. This method is less cumbersome than it sounds, and is useful technique to bear in mind for occasional difficult specimens.

*Transmitted illumination* is, however, the most widely used. Light rays are passed through the object towards the observer. In this way he is seeing an image that is made visible by selective absorption of light—it is, in effect, a shadow of sorts—instead of by reflection. In the conventional microscope, light is admitted to the substage area by reflection from a pivoted mirror. It passes through the object, is collected by the objective lens, and the image formed as a result is examined with the eyepiece.

A great many users of microscopes take the subject no further (and are, one suspects, greatly mortified by their consistent failure to get good results, that is the inevitable consequence). But it is not so simple.

## TYPES OF ILLUMINANT

There are, to begin with, many alternative sources of artificial light for microscopy. The earliest, as we have seen, were candle and oil-lamp flames, later superseded by oxy-hydrogen and incandescent mantle lamps. The arc light was used occasionally, and limelight was employed from time to time (this being the incandescence of lime when heated in a gas/air flame, a kind of illumination sometimes used in theatres before electric lighting became general). By the 1920s the miniature arc of the Pointolite lamp was becoming popular, and it still finds favour in some laboratories, but the mainstay of microscopy has become the filament lamp.

Filament lamps range from the conventional pearl domestic

Page 103 An elaborate microscope—the Burch reflector in use at the Chester Beatty Institute, London. Note ultra-violet source (foreground) and spectrometer (on bench) set up for studies on spectra of haemoglobin. The heavy mirrors are housed in the black metallic housings visible on the microscope itself (upper centre)

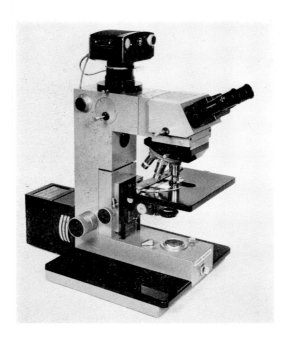

Page 104 (*top*) A research microscope in action. The Vickers Photoplan equipped for transmission photomicrography. Note focussing stage, distal limb of box-like construction, and the low level of the controls

(*bottom*) Specimen assessment may be automated by the use of a scanning microscope used in conjunction with a computer, as in this Zeiss equipment. it is possible to obtain routine measurements of particle size, density and distribution etc on an automated basis. There are possibilities for the screening of cell smears (eg from the human cervical mucosa) in preventive medicine

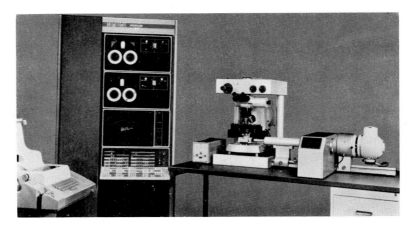

lamp to the high-efficiency, short-life projector bulb. The filament itself may be straight, coiled, or coiled-coil; and the shape of the cap and the bulb itself varies with its design intention. High-voltage bulbs have disadvantages: they are (i) very hot when working and (ii) tend to arc across from one coil to the next.' For this reason the low-voltage lamp fitted with a transformer (often continuously variable, to aid selection of brightness and colour temperature) is the most popular.

Ribbon filaments, which are flat strips of incandescent conductor, are used in some designs; they may be mounted horizontally or vertically, according to the lamp design, but must not be used in the incorrect orientation or the bulb will soon burn out. The mounting of filament lamps is similarly important. These are made so that the filament support and the material from which the cap and gas seal is made are subjected to a calculated temperature, and the burning of a lamp stated to be 'cap up' in the 'cap down' position will incur risks of overheating and premature failure.

Lamps for microscopy are often fitted with a specially-designed filament that has been 'folded' back on itself several times, and in this way shows a bright square or rectangle of relatively even illumination. The wire of the filament is very tightly coiled in the low-voltage (say, 6-12V) lamps; indeed the coil lumen need not be very much greater in diameter than the thickness of the wire from which it is made! This, of course, makes the filament quite robust. In recent years the so-called 'solid source' lamps, in which the filaments are in lateral contact in order to form an even light source of constant dimensions, have become popular. The design was originated by the British company of AEI, but other nations have since produced versions of their own.

There are also many quartz-halogen lamps, originally intended for projectors but now used for microscope lamps and also in motor headlight units, which have a very high output of light with a relatively low temperature and give very good colour temperatures for micrography. Care has to be taken, however, to match the lamp to the current that is designed for it. A lamp burned at 20 per cent under its rated voltage will give only half the light output it should, though it will last ten times longer than its design specification.

On the other hand, an overloading of 20 per cent will roughly double the light output, but will reduce the life of the lamp to one-tenth, or less, of the anticipated time. So voltage regulators are often used. Furthermore, with some lamps of low thermal capacity (such as the ribbon filament lamp), the light may change in intensity with the AC frequency—a fact that becomes important where micrography is the aim, and exposures of $1/125$ second or shorter are to be used.

It is also important to ensure that the cable between the transformer and the lamp is kept as short as possible, the length from the socket to the transformer being just as long as necessary. This is due to the fact that the 240V mains supply is quite enough to lose only a negligible amount between socket and light unit—but a 6V lamp, for instance, will *firstly* show marked colour changes with a potential drop of, say, even 0.25V (which would be inconceivably small at a potential difference of 240V) and *secondly* the voltage is not enough to overcome even the normal resistance of electrical cable for any appreciable distance. Circuit resistance of the order of only hundredths of an ohm can be sufficient to cause deterioration.

*Arc lamps* are used for more specialised applications. The carbon arc is the oldest form, though it is not popular. Probably this stems from prejudice about the difficulty attached to the use of the arc—but modern designs are available in which the carbons may be advanced automatically, thus allowing the arc that has been struck to remain stable for long periods, and if care is taken to thoroughly dessicate the carbon rods before they are used there will be no trouble with the crackling, semi-explosive effects of moisture in them. Baking in a drying oven and subsequent storage in the dessicator are recommended as a rule. This provides one of the brightest sources available to the micrographer—and to be frank it is not usually needed.

The tungsten arc (such as the Pointolite lamp referred to earlier and the various equivalents that have been produced) gives a bright and still popular source of light. Problems arise because of the change in position of the arc and, in some models, the fall-off in quality with time may be a significant drawback. But the tungsten arc is the favourite of many microscopists, and will remain so for many years. The Pointolite system is based on an arc that is automatically struck between a small tungsten ball (the anode) and a filament situated nearby. It has wide-ranging applications in microscopy.

Mercury arcs are popular, if specialised; they give strong output in specific narrow bands in the green and ultra-violet. The monochromatic green at a wavelength of 0.456 $\mu m$ should be more widely used. It gives a very marked increase in image quality, without any chromatism, and is invaluable for critical observations. A change of filter allows the U/V output to be utilised for fluorescence or high-resolution work.

Other arc lamps include the very bright xenon arc, with its colour temperature of around 6,000°K, and the zirconium arc which is rated at 3,200°K and is therefore particularly suitable for type B colour films.

*Light colour* is a question that is dealt with in the host of manuals on photography, and the details of the subject are not of immediate concern to the microscopist. For all normal purposes, glass or gelatine filters are readily available in a vast range of colours and densities. Specific wavelengths can be eliminated, or selected; and the properties of the illuminant can be modified for the work in hand without any great difficulty.

A neutral grey background colour is necessary for the interpretation of stain colour, but it is not as critical as is frequently taught. The eye adjusts very easily to slight colour casts, and this may easily be seen by projecting two transparencies in sequence on to a screen, when they were taken on different makes of film and do not have the same colour balance. At the moment of changing from one slide to another, the difference may be very noticeable to the eye: the second slide may seem to be, for example, all-over blue at the moment it first appears on the screen.

But within a matter of seconds the brain's interpretation of the visual stimulus begins to adjust and reorientate, restoring a normal appearance to it. It is important to realise the arbitrariness of interpretation in this way. The brain is conditioned by experience. It can adjust to very unusual circumstances—which may never occur naturally—and interpret them successfully. In our past obsession with 'optical illusions' we have lost sight of the fact that all sight is in essence an illusion anyway; and the brain is more easily tricked into 'seeing' normality where none exists, than it is confused. We should realise that the age-old 'puzzle' of the reversing staircase is very much less astounding than the ability of the eye to 'see' a face—expression and all—in the few dots of a printed half-tone photograph, or the idle sketch of a doodler. This question of visual interpretation in connection with the examination of micrographs is frequently underestimated or ignored.

So a colour cast may be ignored by the eye to a profound extent. But its effects can markedly affect the appearance of a stained specimen. The use of light that is a little too blue, for example, will darken the red stains (such as safranin or eosin) to an extent that may vitally alter the appearance of the specimen, and even prevent adequate interpretation. Similarly, light that is too yellow will tend to blacken crystal violet or any other purplish stain (nuclei stained with haematoxylin may be affected, for instance) until the structure within the stained specimen is obscured.

So the colour of the illuminant can too easily become a matter of pedantry and argument—but at the same time, the eye's considerable abilities to adapt will not prevent deterioration in the specimen's appearance if the microscopist takes this too far.

For most purposes, no filter is used. If the light from a filament lamp is a little too yellow, a pale blue correction filter may be employed, and will very likely be supplied with the lamp in the first place. When a built-in illuminator is a feature of a research microscope, filters are usually included in the equipment supplied.

Control over stain intensity—of particular use for black and white (b/w) photography—is carried out according to the following rule:

To increase the contrast of a stained feature, select a filter with the complementary colour: but to reduce the contrast of a coloured specimen, select a filter with a similar colour.

For normal use, the colours to bear in mind are those listed in the table below. These will darken—in many cases, almost to the point of total blackness—specimens or specimen areas that are of the listed colours.

| Colour to be darkened (ie increased in contrast) | Colour of filter (similar colour will normally suffice) |
| --- | --- |
| Blue (aniline blue, brilliant cresyl blue, toluidine blue, methylene blue, nile blue etc) | Orange |
| Green (methyl green, malachite green, brilliant green, fast green, janus green etc) | Magenta |
| Yellow (picric acid, martius yellow etc) | Blue |
| Red (eosin, safranin, fuchsin, carmine, orcein etc) | Cyan |
| Purplish (crystal violet, indigo-carmine, thionin, nigrosin etc) | Green |
| Violet (methyl violet, cresyl violet, methylene violet etc) | Yellow |
| Brown (insect chitin, bismark brown etc) | Blue |

For many applications the use of a filter of a colour similar to that of the specimen is helpful in revealing structural details that will otherwise be obscured. Insect chitin, for instance, can be greatly reduced in contrast by the use of a red filter. It can be rendered effectively colourless, if need be, by the use of infra-red photography, qv.

It is important to bear in mind that there are other filters available, apart from gelatine and glass. Indeed many of the glasses used for filter manufacture are hardly 'glass' at all—some contain no silica, for instance, and may be liable to fracture, heat crazing, or even partial solubility in water. But the problem of finding a filter of the desired colour, or occasionally of finding a dealer who can supply one, can be greatly reduced if we remember that liquid filters can be made from dye solutions, or solutions of other coloured chemical compounds, held in parallel-sided containers. They have the advantage of being infinitely variable in terms of strength of colour (up to the point of saturation), and the colour itself can be controlled if desired.

Basically, any coloured solute can be used to make such a liquid filter. The solution is made up to the desired intensity of colour, and poured into a plastic trough or even a glass museum jar (with plate-polished sides). It is helpful to realise that the colour of the filter can be controlled very simply: the degree of colour saturation is a function of

(a) the thickness of the filter container,
(b) the mass of solute dissolved in the filter fluid.

For regular use, cells made for colorimetric analysis (they are 1cm thick) may be obtained, and certainly provide the basis for a system that can be calibrated and which is repeatable.

The most useful of all liquid filters is probably a dilute solution of copper sulphate, $CuSO_4.5H_2O$ in distilled water. It has the dual advantages of correcting yellowness in the beam of light from a filament lamp, and absorbs heat. Indeed, any filter chamber containing pure (ie boiled) water—water which will not liberate bubbles of dissolved air—is an effective barrier to heat radiation from the lamp. It is interesting to note that liquid filters were also the first coloured filters used by microscopists, when coloured glass or gelatine was too difficult to prepare with optical precision.

The more useful solutes for liquid filters are as follows:

*Ultra-violet-absorbing* filters may be made from quinine bisulphate. A 1% solution in water has long been a popular choice. The problem of using this material is that it has a tendency to fluoresce, but the amount of light given off as a result is negligible. It cuts off ultra-violet at the wavelength of 0.4 µm and above. Sodium nitrite is a useful alternative. Solutions above, say, 2-5% cut off transmission at about the same point. The strength of the solution is in no way critical—even a 1% concentration will do, though solutions containing nearer 10% are perhaps more widely employed. Picric acid, neutralised with NaOH solution, can be used; but of the coloured solutions potassium or sodium dichromate are perhaps the most useful. Sodium dichromate in even minute concentrations (of the order of 0.1% or less) cuts ultra-violet effectively, and increasing concentrations give yellow (1.0%) and orange (100%) filters, the colour of which can be selected at will.

*Infra-red-absorbing filters* are usually either water, as described above, or a dilute solution of copper sulphate. A concentration of around 5% $CuSO_4.5H_2O$ is very useful. The saturated solution contains 25% of the salt, but a deeper cell can be used if, for some reason, a strongly coloured cyan filter is required. For photographic use, incidentally, a little methylene blue—not more than 1:100,000—in a 10% solution of copper sulphate will give a virtually complete elimination of all red bands from the spectrum.

*Filters of other colours* are made easily enough. A deep blue, for example, may be made by adding 25% solutions of $CuSO_4.5H_2O$ and $NH_4OH$ in equal amounts, adding a slight excess of the copper sulphate to give a final ratio of 4:3. Alternatively, organic dyes—such as methylene blue—may be employed instead. The solution should be made slightly alkaline if this particular compound is used.

Green filters may be made from solutions of nickel salts, or copper sulphate can again be used as a saturated solution with the addition of 0.5% w/v sodium or potassium dichromate.

Alternatively, a strong (20% $CuSO_4.5H_2O$) copper sulphate solution can be used with an equal volume of a solution containing:

    0.5% tartrazine
    0.003% fast green FCF
    thymol as preservative

to give a good green filter.

Red filters are best made from dyes, eosin being the most popular. It will, of course, successfully control the colour of a section over-stained with eosin if used in sufficient concentration to make a pink filter. More than this and the colour of the section will virtually disappear altogether.

The compounds that produce an acceptable Yellow or Orange filter are discussed earlier (p111). Solutions of picric acid, sodium/dichromate etc are useful for this purpose.

Other colours may be selected from laboratory dyes or stains, such as those listed on p109.

## THE MIRROR

Unless the lamp is situated directly beneath the microscope and in line with its optical axis, some form of reflecting surface is necessary to direct the beam of light through the optical system. A built-in lamp is found in many student's microscopes these days, and is popular in research microscopes. This is good in terms of practicality and ease of manipulation of the microscope as a whole, but it inevitably makes for a certain lack of flexibility in using the instrument for a wide range of purposes. Many of these designs do not even have a facility for the fitting of a mirror if one is required for a particular purpose—and that is a decided disadvantage.

For some applications (the metallurgical microscope being the obvious example) a built-in lamp is fitted adjacent to a prism or set of prisms that reflect the light beam in the desired direction. But the mainstay of the designer's armoury has always been the mirror. As we have seen, it dates from the early days of microscopy, and was fitted to many microscopes as a matter of routine from the early 1700s onwards, and is with us yet—but the quality of many microscope mirrors, and the design of their mounts, is often unsatisfactory.

The mirror should be movable. It should be centerable. And its reflecting surface should pivot about a fulcrum that lies directly on the optical axis. In practice it never seems to do so. The mirror is designed as a purely mechanical device, and not as an optical one at all: the result is that the frame of the mirror unit is simply mounted between a pair of pivots that are attached to the centre of the metallic holder—a functional place for them in purely mechanistic terms, in other words. This makes the accurate alignment of the mirror into a major task of intricate adjustment which most microscopists will neglect as a matter of course.

The mirror is, as a rule, two-sided. A plane mirror is fitted on one side, and a carefully calculated concave reflector on the other. The concave reflector is meant for use without a substage condenser—ie with light from a pearl bulb, or perhaps daylight, illuminating a very low-power specimen (examined with, say, a x5 objective). It should never be used with the condenser in position, for it upsets the detailed standards of calibration that have been built into it. But this guideline is often neglected, too. Without doubt a measure of improvement could be brought about by cutting down the double reflections that inevitably arise from (a) the silvered surface and (b) the glass/air interface.

The most efficient way of doing this would be by the use of a prism—a device which is, however, not used as frequently as it ought to be. Perhaps because the conditioning in designers' attitudes over the past few centuries of development has produced this mirror-fixation... but failing this, a mirror silvered on the upper surface would avoid the optical difficulties of a conventional glass mirror. Probably the easiest way to obtain one would be to purchase a spare mirror for your particular microscope, and take it to be aluminised. It is very likely that the physics department at a local college or university will know where this service could be obtained. This mirror could be used, with care, the old one being retained for routine or undemanding work. If the aluminised upper surface of the mirror becomes pitted or marked by injudicious polishing, the surface can be removed with a little warm $Na_2CO_3$ solution (which dissolves the aluminium according to the formula below).

$$2Al + Na_2CO_3 + 7H_2O = 3H_2 + 2NaAl(OH)_4 + CO_2$$

The glass surface is unharmed by this process, and can be re-coated with aluminium easily enough.

## CONDENSER SYSTEM

It will be recalled that the condenser lens—as a system for concentrating light on the specimen—was an early arrival on the microscopical scene. The Wilson screw-barrel models were fitted with one, and in the nineteenth century they became more widespread. But they were still regarded as merely ways of increasing the 'strength of the light' (rather like the spherical flask of water used by Hooke and his contemporaries). Only when Abbé began to work on the design in the 1860s did it become apparent that the condenser was a vital part of the optics of a microscope. And it was several years before microscopists in general came to accept the idea. Many adhered to the belief that the best images were formed with untouched, virgin light falling on to the specimen without the interference of any lens at all.

Abbé's first 'illuminator' appeared in 1873. The very use of this term was suggestive of its purpose; a condenser might be thought of as a means only of 'condensing' light into a small area —but an illuminator clearly did more than that. In both senses, it would throw new light on the subject. Abbé's first design is still the most widely used substage condenser. It consisted of a truncated spherical lens which approached close to the lower surface of the specimen slide, with a biconvex lens underneath it to gather light from the mirror. Oiled to the slide, it can be rated at NA 1.30. Fifteen years later a three-element lens came on to the market, in which the lower lens was replaced by a convex lens with a somewhat flattened upper surface, and between this lens and the (unchanged) upper element was a concavo-convex lens, the concave surface uppermost. This approached NA 1.40, a figure that is not exceeded even by superior designs.

An aplanatic condenser, of three elements but with a parabolic lower surface to the lowest of them, has been manufactured and provides excellent optical performance (though not corrected for colour); and the most superior performance of all comes from the six-element achromatic condensers with two negative lenses in the design. They are only required for high-power micrography in colour, as a rule.

With these last types, the design has taken into account a range of parameters—including objective aperture and light-source distance—which have to be observed by the user if the maximum benefits are to be obtained. For most uses, these simply become irksome extra difficulties, and the conventional Abbé condenser—with a three-element version for high-power work—is quite enough for all normal light-ground work.

The phase condenser is fitted with the phase ring that produces the phase change distinction between transmitted and reference beam, and it is important to be able to centre the image of this ring accurately. So for these purposes, a centerable substage is vital—and yet it is not always present unless specifically requested. The diameter of the phase ring compared to the diameter of the objective aperture ($A_o$) is given by:

$$0.66A_o > d > 0.5A_o$$

and the width of the annulus of the condenser itself is:

$$\tfrac{1}{15}A_o$$

These are empirical standards set down in Zernick's original work, and a narrower band might provide a better image.

A special focussing telescope is often used for observing the relative positions of the condenser annulus as it is adjusted so that its image exactly coincides with the position of the phase ring of the objective.

Dark-ground condensers are used for relatively few investigations in modern times. This is unfortunate since, as we have seen (p90) they are capable of revealing literally submicroscopic features in an object and the image can be greater in contrast than that obtained by any other means. Bacteria, for instance, may be seen as brightly shining, star-like images against a velvet-black background. It is sometimes found that the technique is described as 'darkfield', but this is an inaccurate designation (p168).

The simplest way to obtain a dark-ground illuminator is to dispense with the condenser altogether, and insert a patch-stop that is large enough to occlude direct light as it attempts to enter the objective aperture, but which is able to converge on the object in the form of a conical beam of light. This technique is suitable for very low-power lenses. It is an easy matter, for higher magnifications, to include a patch-stop of this sort below a conventional condenser, in the filter-holding ring which is customarily 3.3cm in diameter. The light then enters the condenser around the periphery of the lower collecting lens and is emitted in a configuration that is not gathered by the objective. For higher magnifications the size of the patch-stop would be so great as to cut the light down to impracticably low levels, and so a dark-ground condenser must be used.

But there is one attractive sophistication of this form of dark ground that is rarely heard of, but is most aesthetic for low-power work. It was first described by Julius Rheinberg shortly before the end of the last century. In this method, the patch-stop is replaced by a coloured filter, and the annulus around it is fitted with a filter of the complementary colour. Thus the patch-stop might be blue, and the area around it yellow. Filters of this sort can sometimes be found in secondhand dealers, but they are no longer manufactured. In use they give (as in this case) a brightly shining yellow specimen against a dark blue background —a very effective way of increasing chromatic, as well as mere visual, contrast. Filters may be made by cementing, for example, a blue gelatine disc to a pale yellow glass filter.

But for high-power or critical work a dark-ground condenser is required. There are many designs, ranging from a stopped paraboloidal condenser which is useful for lower powers, to very high-aperture internal-reflecting glass elements up to NA 1.40 or even slightly higher. The most widely used are the double-reflecting bicentric condensers in which the incident light is reflected centrifugally by a hemispherical well in the glass element and then, by reflection from the circular outer contour, is concentrated in a conical path that converges on the specimen. Since only a part of the incident light is admitted it would seem that the field is only dimly illuminated. But the concentration of all the admitted light into a small specimen area and the intrinsically high contrast of the technique make this a very acceptable form of illumination. I have been able to take cine film at the highest magnifications with an exposure of 0.1 second, the image showing a level of contrast far higher than could be obtained by any other means. None the less it is important to use a bright light source (such as a projector) for work at the highest magnifications. Some workers have used sunlight, captured through the driven mirror of a heliostat, as illuminant in photographing bacteria at high magnifications.

Oil immersion condensers should be used for critical work. It is not generally realised, but the back-and-forth reflection of light trapped between parallel glass surfaces in the optical system causes a certain amount of glare or 'fog' that degrades the image quality. For dark ground this is a very significant factor, and all dark-ground condensers should be tried oiled to the bottom of the specimen slide, even if a low-power objective is in position. Even a conventional Abbé condenser may show an improvement if oiled to a slide for high-power work. But this is not so much a question of optical correction as it is of the avoidance of any unnecessary glass/air interfaces in the optical system itself. Some modern systems have an auxiliary 'top' lens that swings out of position when low-power objectives are in use. They are dust-traps and must be kept clean.

It should be borne in mind that some modern condensers have been especially produced to give the three forms of illumination —light-ground phase and dark-ground—with only slight adjustments necessary. The addition of a polarising filter gives an additional optical mode, and even Rheinberg filters, $qv$, could be fitted if necessary. Many dyed-in-the-wool microscopists tend to prefer a set of condensers, relying on each to perform the particular function for which it was designed. Other specialised condenser systems have been envisaged in which the two components of the light beam (deviated and undeviated) are separated by a polarisation change, and the optics contain a device that enables the phase or amplitude to be controlled at will. Further developments may be directed towards the use of polarised light in phase systems, which will add colour distinction to phase enhancement.

## *THE SLIDE*

Light emerging from the condenser passes thence through the slide. And, as anyone will readily tell you, slides are nothing more than thin slips of glass, 3 x 1 inches in size, obtained in boxes from stores. A quick polish with a handkerchief, or with the tail of a cotton shirt stealthily withdrawn for the purpose, and they are ready for use ... The intricacies of thickness, grade, pH and the rest do not seem to bother most users at all.

For many purposes, the thickness of the slide is not of great consequence—but for critical work, particularly for dark ground microscopy, it is important to adhere to the limits laid down in the literature that the condenser manufacturer has supplied. Since oil on the under surface of the slide will usefully remove the uncorrected air-space between condenser and slide, there is a margin of adjustment—but to exploit this effectively the use of thin slides (say 0.9mm thickness) is necessary. However, the cleansing of slides is a different matter. Unless they are handled carefully, the presence of slight traces of grease or dust particles (which may be electrostatically attracted to the slide surface) can cause interference with the preparation or observation of specimens.

Alkali in the slide can best be removed by treating it in dilute HCl, followed by rinsing in clean (preferably de-ionised) water and subsequent storage in filtered alcohol—isopropanol is most useful. Alternatively, dusty slides or those that are highly alkaline (and all slides that are to be used for critical interference or dark-ground work) should be boiled in tap-water, steeped in fuming $H_2SO_4$ for *not more than* three minutes (much longer may result in the surface being etched away in parts, causing image deterioration of its own), and then allowed to soak for a few hours in bench sulphuric before being rinsed in filtered running water for *at least* half an hour. The slides may then be given a final rinse in distilled water before being (a) oven-dried or (b) stored in filtered alcohol as before. Slides may be polished with a non-linting cloth. In this case a few fibres here and there are really of no great consequence, but it *is* important to ensure that the cloth has been recently boiled and laundered so that it is quite free of grease. If the cloth is applied to a slide injudiciously, then traces of epithelial oil that were left behind last time the cloth was handled can be rubbed over the surface, causing a complete breakdown of any high surface-tension liquid film that may be subsequently spread. It is best to have several cloths, each with one side marked clearly so that one surface—and only one—is used to polish the slides, handling being confined exclusively to the other surface. After being dried, slides may be treated with a flexible adhesive that may be spread over the surface of the slide or, if a thinner solution is used, into which they may be dipped before being allowed to dry off. The plastic envelope can then be peeled off when the slide is required for use, and any traces of linting or dust deposition will be brought away with it.

Slides are usually available in several sizes, 3 x 1 inch, 3 x 1½ inches, and 3 x 2 inches. These are now designated by the metric equivalents, based on 25 x 75mm and so on. The available thicknesses are shown in the table:

| NUMBER: | 01 | 02 | 03 |
|---|---|---|---|
| THICKNESS: | 0.8/1.0mm | 1.0/1.2mm | 1.2/1.5mm |

The thin slides (01)—that is to say 'number one'—are most useful for critical work, but the increased robustness of the 02 slides makes them best for routine use.

## COVERSLIPS

For some years a situation notable for its patent idiocy existed over the production of coverslips. They were produced in three groups of thickness, coded (as for slides) 1, 2 & 3. The thickness in terms of millimetres is shown in the table:

| NUMBER: | 1 | 2 | 3 |
|---|---|---|---|
| THICKNESS (mm): | 0.15 | 0.2 | 0.25 |
| ,, (in): | 0.006 | 0.008 | 0.01 |

NOTE: Unlike the specifications for slides, which show the limits permitted in each grade, coverslip dimensions are the mean of each group. Variations of up to 20% are expected, and actual selection of any given thickness requires the use of a micrometer screw-gauge.

The element of the ludicrous was introduced by the standards adopted by the producers of objective systems. In Britain and Europe, the manufacturers produced optics corrected for 0.17mm coverslips, whilst in the United States the thickness was 'assumed' to be 0.18mm instead. This is not a great difference—*except that none of the grades of coverslip then available were of this thickness.*

An additional grade has since been introduced, designated No 1½, appropriately enough; it comes between the old grades No 1 and No 2, and is defined as follows:

Coverslip No 1½: thickness mean=0.175mm (0.007in).

This gives a thickness for which the microscope manufacturers have designed—at last—and should always be selected for routine use. Specimens, selected by the use of a micrometer, can be kept for critical purposes but it is my opinion that the increase in image quality that results is negligible. No 1½ coverslips can be used as they come from the box with impunity, I believe.

*The refractive index of coverslips* has been assumed by most manufacturers to be roughly 1.523 (for the sodium line at 0.589 µm) and coverslips manufactured to Chance standards come closest to this in practice. Variations in the second place of decimals have been found in coverslips from other (sometimes anonymous) sources.

### IMMERSION OIL

Where it is used, the beam of light passes from the coverslip into the immersion oil. The optical purpose of the oil is to continue the structure of the spherical front element of the objective, so that the specimen may be considered to lie at the point within the sphere where an aplanatic image is produced. But the picture is not quite as simple as this. Many workers imagine that the whole objective front element/oil/coverslip/mountant complex is a composite structure with a common refractive index. It is not. The first homogeneous immersion oil chosen was cedar wood oil, selected by Abbé after he and Dr Töpel had examined over 300 different oils at the Zeiss establishment in Jena and in the town's university. A system using thin Canada balsam as the immersion fluid had been evolved a few years earlier, in 1873, by the US microscopist Robert Tolles. This was far closer to the truly homogeneous system, for the refractive index (RI) of balsam is close to that of glass. However, cedar wood oil has a lower RI (of around 1.515) and even when the lighter fractions are allowed to evaporate, so that the oil becomes viscous, it does not exceed the level of RI=1.520 or thereabouts. Crown glass rates around 1.524. So the indices of cedar wood oil and glass are never identical.

This is an interesting comment on the generally accepted view, for it shows that the system is not—strictly speaking—homogeneous at all. Lenses are formulated for a given thickness of coverslip at the known refractive index, and take account of the disparity between that and the RI of immersion oil. Even synthetic oils more recently available, and now widely used in microscopical laboratories the world over, are usually rated at an RI of 1.515. Thus it is *not* true to say that a stained preparation may be examined as well by oil-immersion microscopy as may an unmounted object, with nothing but oil between it and the front component of the objective. The objective is corrected for immersion oil in which RI=1.515, and a coverslip thickness of 0.17-0.18mm. *Image deterioration can become apparent at the highest magnification even if a No 1 or No 2 coverslip is used in place of No $1\frac{1}{2}$—and an unmounted specimen will give a significantly degraded image.*

However, a certain degree of correction can be carried out by adjusting the tube length of the microscope, if this is possible, on a given instrument. The tube lengths that allow for coverslips of a range of thicknesses are shown in the following table. There are the approximate figures for a $\frac{1}{6}$in (4mm) objective which is used (without immersion oil) and where the differences are therefore more significant.

| COVERSLIP NUMBER | THICKNESS OF COVERSLIP | | DRAWTUBE LENGTH: |
|---|---|---|---|
| | mm | in | |
| | 0.10 | 0.004 | 240mm |
| | 0.135 | 0.005 | 200mm |
| No 1 | 0.15 | 0.006 | 180mm |
| | ⎱0.17 | 0.007 | 160mm |
| No $1\frac{1}{2}$ | ⎰0.19 | 0.0074 | 150mm |
| No 2 | 0.20 | 0.008 | 140mm |
| No 3 | 0.25 | 0.010 | 130mm |

As a rule of thumb, it should be borne in mind that a *thin* coverslip may be corrected by using a *greater* tube length; conversely a *thick* coverslip may be corrected by a *lesser* tube length. If a batch of coverslips is suspect, it is a simple matter to adjust tube length by, say, 5mm at a time until a more acceptable image is obtained.

It is also useful to realise that the microscope itself may be used to gauge the thickness of a coverslip in certain circumstances. This depends on a fine-adjustment in µm. The measurement is undertaken by cleaning the slide and coverslip and observing the presence of interference fringes ('Newton's rings') as an indicator of very close proximity when the coverslip is placed on the slide. It should then be possible to focus:
(a) on the slide surface adjacent to the coverslip; and
(b) on the upper surface of the coverslip itself.
The calibration markings are noted on each occasion and the difference is the coverslip thickness.

An alternative method is to image the lower surface of the coverslip through the thickness of the coverslip—but this will not give a direct reading as the refractive power of the glass will give a result that is too low. The result should be multiplied as follows to give the correct thickness:

$$d_c = 1.5 \, d_a$$ where $d_c$ = coverslip thickness
and $d_a$ = apparent reading obtained through glass

It is equally possible, of course, to calibrate the fine adjustment of a microscope over a given range and in this way use it as a 'depth-finder' by referring to a small drop of paint or a scratch mark put for the purpose near the periphery of the control knob.

Different immersion fluids may be required for special purposes. Several compounds have a refractive index near that of immersion oil (RI = 1.515), including: benzene, xylene, crown oil, sandalwood oil, and hexachloropropane.

For certain purposes the solvent action of immersion liquids or the nature of an immersion oil makes an aqueous immersion liquid preferable. This has been recommended in the past, for example, for use in interference microscopy. Sodium iodide is soluble in gycerine. A very small amount is enough to dissolve the crystals—a 5% addition of glycerine will provide a viscous material with an RI of roughly 1.52. The addition of further glycerine (RI=1.472) will reduce this to the level required. Glycerine is hygroscopic, however, and this mixture will absorb water and become somewhat lower in refractive index. For long-term use it is better to dilute the 20:1 NaI:glycerine mixture with concentrated glucose syrup until the desired refractive index is obtained.

Special oils are also required for fluorescence work, as many immersion liquids will fluoresce sufficiently to cause image degradation. They may be ordered under proprietary names, but it is worth noting that it has been found that nitrobenzene in small amounts may be added to an immersion oil with the effect of greatly cutting the fluorescence. Certainly this would be worth trying before any more sophisticated answer was sought. Note that the mixture will not be stable for prolonged periods as it will tend to darken; a fresh batch should be made up if this occurs.

Finally we should remember that water and glycerine have both been used in the past as non-homogeneous immersion liquids. They are, of course, quite useless for a modern highly-corrected optical system. But of course, if a water-immersion objective is to be used (and a few are still found) oil would be similarly inappropriate.

## OBJECTIVES

### Achromatic objectives

The greatest percentage of objectives in use is achromatic. These are descendants of the lenses designed in the nineteenth century, and are chromatically corrected for two colours, whilst being spherically corrected for one wavelength near the middle of the visible spectrum (in the yellow-green). They give adequate performance for normal use, though a yellow-green filter will be needed for successful high-definition photography in black and white (b/w). For colour micrographs, a filter to cut ultra-violet may be necessary. The achromats most widely available are as follows:

| DESIG-NATION | FOCAL LENGTH: | NUMERICAL APERTURE | MAGNIFI-CATION | FIELD DIAMETER |
|---|---|---|---|---|
| 2in | 50mm | 0.10 | x 2.5 | 10mm |
| 1½in | 32mm | 0.15 | x 4 | 5mm |
| ⅔in | 16mm | 0.28 | x 10 | 1.8mm |
| ⅓in | 8mm | 0.50 | x 18 | 0.75mm |
| ⅙in | 4mm | 0.85 | x 45 | 0.42mm |
| ⅛in (O.I.)* | 3mm | 0.95 | x 60 | 0.3mm |
| 1/12in (O.I.)* | 2mm | 1.30 | x 90 | 0.18mm |

*O.I. signifies an oil-immersion objective.

The lens array is mounted in a component that screws into the end of the objective housing, and may be removed *in toto* for careful cleansing. *On no account* should any attempt be made to take the lens housing down or remove the lenses. Dirt between the lens elements is the manufacturer's responsibility: it was either there to begin with (in which case the lenses should be returned because of their failure to meet specifications) or it has gained admittance through some production weakness which is a manufacturer's failure too). In practice, as one would expect, neither of these failures is ever seen: if dirt gets in, it invariably does so as a result of misguided attempts to take the array to pieces. As we have seen, some objectives of the past were made in such a way that the only means of seeing their construction would be to saw them in half from top to bottom . . . and modern objectives are designed as a dustproof, not-to-be-tampered-with unit. It is important to teach this.

*Flat-field ('plano') objectives*
As the human eye looks at a scene, the only part that is actually studied closely—which is 'seen clearly' as we would say—is in the centre of the field of vision. The rest is a relatively indistinctly visualised area which serves to place the object of attention in its visual context.

And so it is with a microscope objective. The eye (quite instinctively) looks at the near centre of the circular field of view, and looks at it intently. The rest of the circle is not studied closely, or if some object is there to which it is desired to give special attention, it is instinctive to move it into the centre. For this reason the achromatic objective has a very sharp and clear central definition, but this falls off in the periphery of the field of view. When the eye is presented with a micrograph, of course, it scans the whole area in order to locate information. And at once a serious drawback becomes apparent.

Only the centre of the photomicrograph is clear, if it was taken with a conventional achromat. Though this is of no consequence when the 'centre' is elective, ie when the slide can be moved so that any area is scrutinised, this is clearly impossible with the photographic image. Only the original centre is in focus, and the rest is—to a greater or lesser extent—slightly indistinct.

To compensate for this phenomenon, flat-field objectives have been designed. The principle of construction is founded on the use of thick lenses, corresponding to the *central portions only* of larger lenses, as it were. In many designs the compounded lenses are, if we may borrow a term from motoring, oversquare; ie they are taller than they are broad. The flat-field correction gives a very even coverage, without peripheral fall-off in quality, but it is attained at the expense of a slight loss of overall definition. Representative ratings for flat-field objectives are given in the table:

| DESIG-NATION | FOCAL LENGTH | NUMERICAL APERTURE | MAGNIFI-CATION | FIELD DIAMETER |
|---|---|---|---|---|
| $1\frac{1}{2}$in | 35mm | 0.09 | x 4 | 5mm |
| $\frac{2}{3}$in | 16mm | 0.25 | x 10 | 2mm |
| $\frac{1}{3}$in | 8mm | 0.50 | x 18.5 | 0.9mm |
| $\frac{1}{6}$in | 4mm | 0.65 | x 40 | 0.45mm |
| $\frac{1}{12}$in | 2mm | 1.25 | x 100 | 0.20mm |

For many designers the easy way out of the many technical problems involved in making a flat-field objective has been to leave some residual colour in the image formed by the objective, correcting for this by specially produced compensation eyepieces. Their use without the flat-field objectives (or the use of objectives with the conventional eyepiece) would result in an unacceptably chromatic image. This error is sometimes made by microscopists quite inadvertently.

There are also apochromatic flat-field objectives, with a higher numerical aperture. They are amongst the most intricate and expensive of microscope lenses; but the numerical aperture of a x 100 oil immersion objective made in this way may be as high as 1.32, compared with 1.25 for the equivalent achromat. For the most critical micrography such a lens is useful; but it is superfuous otherwise. Conventional apochromats are discussed more fully in a following section.

*Fluorite objectives*
These lenses are not to be confused with apochromats, *qv*. They are so named because the earliest versions were made with natural fluorite, a mineral often used in apochromatic lens systems, but now the term is applied to lenses with corrections superior to achromats but not as good as the apochromatic equivalents. Many feature an element of synthetic fluorspar. The cost of these lenses is a little more than that of achromats, but the increase in quality is probably worth while. Indeed they are a better buy for almost every high-power application than the equivalent apochromat, which gives only a slight further increase —but at a very much greater cost. For example, the NA 1.30 x 100 lens is a very valuable addition to any microscope outfit.

*Apochromats*
These are expensive, technically precise lenses of considerable complexity and sophistication, corrected for chromatism at three points on the spectrum and for spherical aberration in two colours. They give the best image quality in these terms that can be obtained. The NA of a x60 lens can be as high as 1.40,

for example; though this is inferior to some nineteenth-century objectives. There are drawbacks, however. The field of view, though sharper at its centre, falls off towards the periphery more than in many achromats, which makes their use for photomicrography somewhat limited; and they may need specially made compensating eyepieces.

*Other special objectives*

Some objectives are fitted with a correction collar for use with specified coverslip thicknesses; a few are manufactured with an iris diaphragm that can cut down the numerical aperture for use in dark-ground work. The NA here should not exceed 1.0 for normal work, though a special condenser may permit lenses rated as high as 1.2 to be used without permitting field light to enter the objective.

Lenses for polarising microscopes are selected for their strain-free glass components (the optical activity for some minerals may make apochromats unsuitable for such applications). Objectives for phase microscopy, interference microscopy etc are made according to the different demands of manufacturers' systems, and the manual supplied with each one should be studied before they are used.

## THE MICROSCOPE BODY

The primary image formed by the objective is projected through the body of the microscope to the eyepiece. Some years back it would have been a simple matter to dismiss it in a few words as a metal tube with lenses on each end ... but the more sophisticated equipment that many microscopes currently embody has complicated the issue somewhat. Many research instruments have built-in cameras or a choice of viewing/projection facilities, and prisms direct the image rays along their paths. Each design is unique, and there would be no point in attempting to categorise them all. But it is important, once again, to leave well alone. The removal of prisms from a light breathe-and-polish may cause alignment problems or result in the introduction of enough humidity to cause misting, the introduction of dust particles, or inadvertent finger-marks that will cause flare. Prisms exposed during component interchange may be wiped with lens tissue or a non-linting cloth kept for the purpose of polishing lenses and slides (p119).

The interior of the microscope body should be matt black in order to minimise internal reflections. Blackboard paint is useful if any extra retouching is needed, and it is possible to buy black paper of low reflectivity designed for use in cameras. It is always possible to control internal flare of this sort with the body of an eyepiece from which the lenses have been removed. Most eyepieces have a flange somewhere inside which cuts aberrant light, and if the lenses are removed and laid carefully to one side, the lensless eyepiece will serve as a baffle. Often the microscope design does not permit the use of this idea, as the baffle slides down too far and cuts off the periphery of the field of view. But trial and error will show whether there are applications for this principle in any particular case.

At the lower end of the body is the objective holder. Into this the objectives are fitted. The screw thread that has been accepted as the international standard for all objectives (much to the confusion of the metrication movement) is a whitworth thread, that is to say a v-shaped thread in which the sides in transverse section are inclined at an angle of 55° to each other. One-sixth of the depth of the v is rounded off at the top and bottom of the thread; but for the US version these amounts of truncation are slightly greater. The pitch of the thread is 36 to the inch (0.7056mm) and the length of the threaded portion is 0.125in (3.175mm); its (greater) diameter is 0.7982in. The objectives screw properly home, shoulder-to-shoulder.

This specification was originally laid down by the Royal Microscopical Society, as were several other useful parameters (such as eyepiece diameter etc). Sadly, the others have not received international acceptance, and so though any objective is likely to fit most professional microscopes, interchangeability of other components is more difficult. This can be very annoying, particularly when different manufacturers produce the ideal objective, condenser, and eyepiece for a particular task, some of which do not fit the stand available.

There are several types of objective holder available. The most simple is the simple threaded nosepiece, on to which a single objective can be screwed. There are more complicated versions, such as the centerable lens changers that may be found, but the most widespread holder by far is the rotating nosepiece holding three, four or even five objectives. This locates to a spring-loaded stop mechanism which ensures alignment after a lens has been changed.

At the other end of the body, of course, lie the eyepieces. They are held in the head of the microscope, the simplest form of which is a single, vertical tube. More sophisticated than this is the inclined head, which allows the observer to keep the microscope upright whilst viewing at a restful angle. Similar in principle is the binocular version, which is invariably angled in this way. There are also 'trinocular' heads in which an extra, vertical tube is available for microprojection, camera attachments or whatever.

With many modern microscopes the tube length is fixed. Usually it is 160-170mm, but in some metallurgical microscopes it may be well over 200mm. Often, even if it is not possible to correct tube length in the ordinary way, a monocular head is available that can be so adjusted if need be.

As we have seen, there are many reasons to adjust tube length. The presence of such a facility should be borne in mind when selecting apparatus.

## *EYEPIECES*

The normal microscope eyepiece is the oldest manufactured component in the history of the instrument. It was around 1685 that Christian Huygens designed a viewing lens system which consisted of a pair of lenses, both plano-convex, and mounted at ends of a short tube so that the flat surfaces of both elements faced in the same direction, that is upwards or towards the observer. Huygens noted that the strength of the eye lens should be three times that of the field lens. And this system is by far the most widely used eyepiece in modern microscopy. The conventional Huygenian eyepiece is made in a range of magnifications from x6 to x15; the x8 or x10 are often used for routine work. Selected focal lengths are as follows:

| FOCAL LENGTH | MAGNIFYING POWER |
| --- | --- |
| 42mm | x 6 |
| 25mm | x10 |
| 17mm | x15 |

The original Royal Microscopical Society standard for objective dimensions set down an external diameter such as to fit into a drawtube of 0.917in internal diameter. The magnification is normally quoted as it applies to an image distance of 10in (250mm); long held as being the distance of 'optimum vision'—an assertion with which many would disagree.

The numerical aperture of the eyepiece is more or less equivalent to that of the objective divided by the magnification and so it is small; similarly the magnification of the eyepiece is low. The compensating eyepieces that are needed for certain objective types (*qv.*) are produced in a wider range that includes some more powerful lenses:

| FOCAL LENGTH | MAGNIFYING POWER |
| --- | --- |
| 45mm | x 6 |
| 30mm | x 8 |
| 25mm | x10 |
| 15mm | x17 |
| 10mm | x25 |

A third range is produced under the designation flat-field eyepieces. They give an excellent flat field when used with the modern objectives (p125) and are best suited to binocular microscopes.

## THE STAND

Microscope stands are now available in a bewildering variety. The foot itself is quite often based on the horseshoe shape of the 1890s, and this is practical for many reasons. An illumination base can be easily fitted, and for specialist work, a mirror mounted on a baseplate can be substituted for the built-in reflector if this aids centerisation.

Many microscope stands these days are based on the box principle; there being a flat rectangular box for the base and a vertical box in place of the more orthodox limb. The controls are grouped in anthropometrically correct positions as far as possible. Larger microscopes have in recent years tended to 'turn round' as we have seen earlier. Whereas the limb was traditionally adjacent to the microscopist, now it is often on the far side, away from him; and the components are literally 'hung' on the front. Within the stand are the focussing components.

### The coarse adjustment

This is usually the largest milled knob on the microscope stand. It is typically connected to a rack-and-pinion mechanism by which the whole body and the lenses attached to it can be raised and lowered. The gear ratio is low, so that a turn or two is enough to take the tube from its lowest to its very highest positions. The teeth of the gear should not be right-angled to the rack, but cut helically (ie at an angle) so that several teeth are engaged at one time. This makes for a smoother movement. There should also be a means of adjusting the controls to compensate for slackness, looseness or backlash that may easily develop. Many microscopes of today have no such facility. This is ostentatiously because 'it would not be necessary' but no doubt adjustments had to be made before the microscope was passed at the factory. Too tight a mechanism will prove tiring and irksome; too loose a movement will allow the body tube to descend when pressed, a fact that can cause damaged slides during micrography.

Many coarse adjustments rely on gravity to eliminate backlash and keep the moving parts in close contact at all times. This can be supplemented by the use of one-way springs to tension the movement—and this is particularly necessary if the microscope is to be used on the horizontal optical bench, eg for photomicrography.

*The fine adjustment*

There are many designs for the fine-adjustment mechanism, which is used to bring the image into final focus after it has been roughly visualised by the use of the coarse control. The mechanism is frequently carried on ball-bearings and is usually sprung (in one direction) against backlash. As a rule, a single rotation of the fine-adjustment control will raise or lower the focus by 0.1mm; the total movement in the mechanism is around 2mm altogether. Probably the most popular traditional fine-adjustment mechanism was the geared lever, acting at one end; but an eccentric cam wheel is frequently used in modern designs. Lubrication is usually unnecessary, but traces of graphite or molybdenum disulphide may be applied if the need arises.

Whatever the system, it is vitally important that fine-adjustment is mechanically sound. It should impart no sideways motion to the image when the direction of rotation is changed; it must not show free movement or backlash either; and it should not be too stiff or too loose in operation. In normal use the control is being continuously adjusted up and down, giving a feeling of depth to the object on the slide, and the amount of resistance in the mechanism is surprisingly critical. As a rule it can be adjusted. In many modern instruments both fine and coarse adjustment knobs are mounted concentrically. Thus the same hand can use each control without moving from one position to another, a most convenient way of working. This is a descendant of the concentric mechanical stage controls which became increasingly popular in the nineteenth century, *qv*.

*The substage controls*
The microscope substage controls vary from type to type. Some have a centerable substage, and these control knobs are mounted at 45°, more or less, to the sagittal section of the microscope body. Between them and the microscope body itself the condenser focussing controls are located. These should enable the condenser to be raised until its upper element is level with the stage surface, and lowered until it can be swung out of the optical axis altogether.

Other controls may be found on special condensers (such as the multi-purpose models described earlier). And there is often, in addition, a facility for the selection of filters, stops or apertures.

In many models the focussing controls as a whole are situated beneath the level of the stage, including the coarse and fine-adjustment. And it is frequently found that the stage is raised and lowered by their rotation, whilst the body tube remains fixed. This is a relatively recent trend. The drawbacks of this system lie in maintaining the firmness and exactly level orientation of the stage, which is normal to the optical axis of the microscope. But several manufacturers have overcome these difficulties and the concept is now well established.

*Special microscope features*
There are many specially designed stands for particular purposes. Some are truly vast constructions housing cameras and scanning equipment. In some microscopes for tissue culture work, the optical pathway is *down* through the specimen; the objectives are mounted underneath the stage, and the eyepieces in a binocular head are situated low on the instrument. Some have novel focussing arrangements: thus there is at least one model in which coarse and fine-focussing are carried out by the same con-

trol knob. A slight backward turn frees the coarse-mechanism and permits fine-focussing; and the same manoeuvre performed in reverse re-engages the coarse-mechanism. This is one of those devices that is something of an acquired taste...

Mechanical stages are now in general use. Even more useful in many applications is a stage that rotates. I find the rotating stage on my own research microscope is of particular value when lining up a specimen for a 35mm colour transparency—when it is important to adequately fill the frame. Sadly, there are slides available (such as the 45mm slide used for petrographical analysis) that do not fit the orthodox mechanical stage.

Another accessory that is useful in fluorescence microscopy is an ultraviolet screen that protects the microscopist from rays diffracted by the specimen. Additionally, if tube length cannot be altered in a given microscope, it is possible to buy correctors that screw into the nosepiece, between it and the objective, and can give an optical correction equivalent to a change in tube length ranging between 100mm and 300mm.

*Special eyepieces*

Eyepieces are available for many special purposes. The crude but widely used pointer eyepiece is simply a Huygenian eyepiece in which a small pointed lever, actuated by a minute handle in the upper surface of the eyepiece, is adjusted to indicate some feature or other in the specimen. They were designed for teaching purposes, but are largely a waste of time. Other eyepieces have polarising analyser filters, or ruled grids for micrometry; and some have been specially designed for photomicrography.

## THE ZOOM ATTACHMENT

Mention must be made of the zoom microscope. This instrument incorporates a special array of lenses, actuated by cams, that provide a continuously variable range of magnifications, perhaps from x1 to x5, with the result that—at the turn of a milled ring around the microscope body—the magnification can be continuously changed from, say, x100 to x500 and back again. There are undeniable applications for this principle, which is of particular assistance to the micrographer who wishes to accurately frame up a shot before impressing the button. But the facility to zoom in and out of the scene, like a trigger-happy football cameraman on a Saturday afternoon television transmission, is possibly overrated. No doubt when the optical quality of the zoom system has been extended until it is clearly comparable to the best of conventional instruments, the zoom microscope will be in every laboratory and we will take them for granted as we now do bunsen burners or bottles of balsam . . . but they are, at present, somewhat superfluous for normal usage.

CHAPTER SIX

# THE MICROSCOPE IN USE

CORRECT AND conscientious adjustment of the microscope, whether it is a simple instrument or one of the highest complexity, is essential if it is to give of its best. That is more than a truism. Many laboratories are content to use microscopes, even very costly microscopes, that have been hastily set up; that are poorly adjusted; or that have dust or some other form of material on component lenses.

It is imperative to cultivate good driving habits in microscopy. With any process—from bakery to high-altitude flying—it is helpful to learn first the tried-and-tested method, whilst being ready to make modifications in the light of personal experiences later on. This is a very different matter from the learning of routines that have been set out under the guise of authority by someone who has never tested the method nor attempted other versions of it. The successful student learns to distinguish between the two: between the arbitrary dictates of an autocrat creating a conformity with which to conform (like, sad to say, much of our educational routine) and the practice that had evolved from practicality, the procedure that has survived because it was fitted to its purpose.

It is arguable that one could simply take a microscope and *use* it—pragmatically working out a procedure that gives results. In the same way one could doubtless drive a car according to an individual scheme, a new approach, if dented bodywork and ground gears were no object. The very poor standard of much modern micrography, even when the most sophisticated apparatus has been used throughout, tends to suggest that it is exactly this trial-and-error approach that has been used. It does not work.

The vast improvement that can be brought about in image quality by slight adjustments, or by removing an eyepiece and wiping it with tissue, is a salutary lesson for many seasoned microscopists. As a test, try this when next you have set up your microscope for use. Whilst observing, rotate the eyepiece, moving it back and forth with the forefinger and thumb of the (left) hand. Or loosen the objective and rotate it slightly too. When the condenser is focussed and high in its mounting, rotate it as well—and observe the movement of the out-of-focus lens markings that had not been noticed previously.

## 'STUDENT' MICROSCOPES

To start with, let us examine the setting up of an uncomplicated microscope of the traditional pattern. Many of the cheaper versions of such 'student' microscopes are now manufactured in Japan, or they may be imported from the Soviet Union or some other East European country which—in order to bring convertible currency into the exporting State—may result in very economical purchase. This does not apply to the basic microscope stand, either; many laboratories are now using research microscopes from these sources. Alternatively, a second-hand microscope may be purchased from a reputable dealer (even from a disreputable one, if you take a close look at it first). What points should the purchaser look for?

Thinness and wear of the enamel or other surface finish on the limb of the instrument near the right-hand coarse adjustment control will indicate that the instrument has been very extensively used in the past, and may point to wear in the mechanism.

It is important to feel for backlash in the controls. Put a slide on the stage (failing that, look at fibres along the torn edge of a piece of paper or card—which is a handy 'test object' when no slide is available) and focus *up* until it is clear. Start with the objective lens close to the object. Then try various changes of focus, moving down to the object, then up to it, altering the direction of movement just as the image is focussed. If there is any play or backlash beware of the instrument as it is, for the play will cause focussing drift during use and blurring of the image in photomicrography. It is a very serious limitation, in other words.

*If there is no backlash* detectable on any of the controls, proceed to examine the rest of the instrument as described below. *If backlash is present*, it is imperative to see if there is a mechanical way of correcting it. There may be a screw attachment that can take up the slack; if not, it is advisable to leave well alone. Perhaps a dealer may be generous in allowing a price concession if the fault is pointed out to him, if you are willing to take a chance on rectifying it.

Examine the inclining joint in the limb, if one is present. It is obvious that the joint must be loose enough for ease of operation, but tight enough to prevent the angle of inclination altering under slight pressure. It is also helpful if the microscope stand will incline through 90° so that the optical axis is parallel to the bench surface—and in this case the microscope should be able to stand firm, the centre of gravity being within the area of the foot. If it tends to fall over backwards, a separate base, with anchoring points, will be required for horizontal microscopy.

Check the lens mounts for damage. A dropped eyepiece may have a cracked lens that may not show during a cursory use of the microscope, but will detract considerably from the results that can later be obtained.

Check for general structural damage. A microscope may have been knocked over, even dropped, and the comparatively minor error of lifting the instrument by its focussing controls (instead of by its limb, as is correct) may have bent a weak shaft. The objective changing mechanism should be operated too, though slackness here is usually a simple matter to rectify.

Finally examine the accessories. Is there a substage condenser?

Does the iris diaphragm work (it is a favourite past-time of student microscopists to take it to pieces, and then fail to get it together again)? What about the mechanical stage? How many eyepieces are there, or spare objectives? And do they look as though they have been taken care of by the previous owner?

It is not necessary to purchase a microscope with a substage condenser at all, to begin with. But it is obviously helpful to bear in mind the possibility of broadening the scope of one's equipment in the future, and so provision for the fitting of a condenser is obviously necessary. One feature it is not possible to dispense with is a fine adjustment. Many student-type microscopes feature a single focussing control which can be either tiring (if the gear ratio is low) or difficult to focus correctly—especially with the high powers—if the ratio is high.

Many microscope cases are fitted with a key, and it is as well to put the spare away (or obtain one, if none is supplied) for the occasion when the case key is mislaid. This is a point all the more important because the case may well have no other means of securing itself shut and it is always irksome to have to find a key for such a simple purpose. A hook closure is worth fitting to a microscope case, if one is not there already.

The microscope may refuse to come out of its case at the first attempt. This will be because of a master screw set into the base of the case which holds the instrument firmly in position. The screw, if present, is best loosened by means of a coin. Once out of its case, it is worth while finding a plastic bag suitable to act as a cover for the microscope when it is in the open.

*Adjusting the microscope*
The basic steps for setting up critical illumination for the bench microscope of the 'student' types are as follows:
(a) Locate the microscope on a firm base and adjust the drawtube length (where one is fitted) to 160mm.
(b) Set up a frosted lamp 9-12in away, preferably in a housing that screens direct light from the observer.

(c) Bring the conventional 'low-power' objective into position (ie the No 3, ⅔in lens, or 16mm) and insert a x8 or x10 eyepiece.

(d) Tilt the substage mirror until the field of view in the eyepiece is flooded with light. If the light is uncomfortably bright, close down the iris diaphragm or choose a smaller aperture in the stop disc (whichever is fitted). If an integral illuminator is fitted, simply turn it on.

(e) Place a permanent preparation on the stage. The best type of slide to choose is a section of some kind—preferably a plant stem such as *Zea mais*. Then lower the objective until it is a few millimetres from the coverslip. Make sure the specimen is in line with the objective.

(f) Looking through the eyepiece, gently raise the body tube (or lower the stage, p134) by rotating the coarse adjustment, until the specimen appears in focus. Move *through* the focal position —ie continue rotating the knob until the specimen has appeared in focus and is *just* beginning to become blurred again.

(g) Fine-adjustment is now done until the image is clear. It is at this stage that it is tempting to continue observations. For critical illumination there are several stages still necessary—but it is interesting to note the appearance of the image at this state, in order to compare it with the appearance finally obtained.

(h) The image of the lamp is now focussed in the field of view by adjusting the condenser controls. This position will be with the condenser high in its mounting, a very small way beneath the slide itself. Alternatively, a seeker or a mounted needle held against the frosted glass surface may be imaged in the eyepiece. When this has been done, *raise* the condenser slightly to render the image of the ground glass *just* out of focus. It will likely cause visual effects of its own otherwise. (It is also possible to slightly *lower* the condenser to obtain the same effect. This is what I always do, but most other workers recommend the procedure outlined above).

(i) Now remove the eyepiece, and close the iris diaphragm until only a pin-point of light is visible. Centre the condenser—if centering controls are fitted. If they are not, this will show how central the alignment is, and some adjustments can be made on an *ad hoc* basis if necessary. But use caution with an old microscope—perhaps it is the nosepiece which is preventing the objectives from lining up true.

(j) Open the diaphragm (or select an aperture in the wheel) until the field of view is roughly three-quarters filled with light. A circular 'frame' around the illuminated aperture should be clearly seen.

(k) Replace the eyepiece. Look again at the field, and if the light seems too bright or dark for the specimen for personal preference, alter it by adjusting the diaphragm or by selecting another aperture stop.

(l) Now remove the specimen and check for marks on the various optical components. Rotating the suspect lens, first the eyepiece, then only its upper lens (by unscrewing the mount slightly) and so on through the system, may enable a dirty or scratched lens to be identified. The condenser may be adjusted to see if dirt on the lamp is being imaged, or rotated to detect dirt, dust or marks on its own upper element. Dirty lenses should be wiped with tissue or cleaned with a cloth *just* moistened with xylol.

(m) Now lower the objective slightly, replace the slide and focus up to obtain a clear image. Test the other lenses in the rotating nosepiece, if one has been fitted. Watch carefully from the side as the lenses are changed, in case the longer high-power objectives will foul the preparation coverslip. Look down the eyepiece and make a cautious adjustment if the image is nearly in focus. If it is, the two objectives so far examined are said to exhibit *parfocality*. If they are not compatible, it is necessary to lower the objective cautiously, as before, and then focus up to obtain a clear image. Remember, however, that the object/lens distance with a high-power lens will be very much less than it was last time. Now change back to the low-power objective and observe which way the adjustment is made. This will give a guide-line as to the amount of rotation necessary to focus the image when the lenses are changed. It is important to note that parfocality can be restored by altering the drawtube setting slightly. To do this, note the degree of rotation necessary to focus the slide after the high-power lens has been replaced by the low-power objective. Then alter the drawtube length by, say, 10mm, and repeat the manoeuvre. If the amount of adjustment is now greater, try altering the drawtube by 20mm in the opposite direction (ie 10mm away from the initial setting) and take a new reading. By this means it is possible to find a tube length conferring parfocality on the objective which, if it is not more than 5 or 10mm from the original setting, will not cause a deterioration in image quality of any significance in high-power use. It is not possible to extend this process to oil-immersion, however, if more than a few millimetres are involved.

(n) To test the oil-immersion objective, rotate the nosepiece to engage it in position and then lower the condenser and swing it out to one side. Place a drop of immersion oil on the plane surface of the upper condenser lens. Then raise the condenser until the drop of oil touches the slide and spreads out.

(o) Raise the body by rotating the coarse adjustment. Place a drop of immersion oil (R.I.$=1.515$) in the middle of the region indicated by the spread condenser oil. Then lower the body until the objective is immersed in the oil, and very nearly touching the coverslip.

(p) By cautiously focussing up, obtain a clear image. Slight adjustments of the condenser are possible to clarify the picture. Care must be taken (a) not to lose the contact between condenser and slide (through the oil droplet) and (b) to keep the iris diaphragm wide open; without this the aperture will be lost and with it the resolving power of the lens system.

(q) A high-power eyepiece (say, x15) may now be inserted instead of the x8 or x10. This will provide the highest magnification obtainable with the greatest capacity for resolution. Higher magnification can be obtained by increasing the drawtube length, but if this is more than 170 or 175mm, image deterioration will be apparent.

(r) In examining selected areas of the slide, care should be taken not to lead oil on to the stage itself from the under-surface of the slide. After use the slide should be lifted—not slid—from its position, and cautiously wiped with a non-linting cloth. It is helpful to moisten the slide with distilled water (breath condensate is the most handy) in order to assist the removal of the traces of oil remaining. The objective lens should be cautiously wiped too, the final traces of oil being removed with a cloth barely moistened with zylene.

(s) After use, engage the low-power objective and the normal (x8 or x10) eyepiece, and replace the plastic cover to prevent dust gaining access to the instrument.

Modifications to this framework of operations are made in special circumstances, of course. But this sequence of events is the seat of microscopical technique and even the most advanced user of sophisticated equipment uses methods that are much the same.

The most important adjustment that has to be made is the purely subjective orientation of the user's mentality. The observer of microscopic objects does not (as usually occurs in films and TV epics) lean casually over his instrument, squint one eye shut, and then give a cursory look down the microscope with the other one... Information is gained by prolonged and careful study, moving the preparation back and forth, altering focus and consigning the received information to memory so that a balanced interpretation can eventually be assembled.

In the mass-media presentations there is a near tradition for the misuse of microscopes by actors who play scientists, the most frequent being the adjustment of the focus so that the oil-immersion objective is four or five inches above the slide. Second most common is the diagnosis of minute detail—white cell differentials, for example—with a binocular low-power instrument giving a magnification of x7 or thereabouts. Third most frequent to catch my eye is the coining of definitive diagnoses after no more than two seconds' observation. Leucocytosis with a shift to the right; strychnine in blood smears; even blood groups and fingerprints feature in these examples, which must leave a lingering envy in the minds of the microscopists who have to be content with less dramatic, if more feasible, investigations as a rule.

The stance of the microscopist is important. A relaxed posture is a prerequisite, the middle forearms resting on the bench surface. The shoulders should be relaxed, and the control knobs are held carefully—not merely 'grasped'—with the thumbs lying along the milled edge in the direction of rotation, the forefinger beneath being crooked at right-angles. The observer must be ready to lean to one side to check the position of the objectives when focussing, and not to restrict himself exclusively to observing through the eyepiece.

The dogma of keeping both eyes open is one of the most widespread of monocular microscopy. It is not necessary. The casual user, indeed, may find it annoying to try to keep both eyes open, whilst mentally concentrating on the view of one eye only. For him, and for the student of medicine or biology, the choice should be left optional. If there is a strong urge to close the unused eye, it should be taken as an instinctive response to a condition of strain. Rather than force the brain to accept what it finds tiresome, and in this instance there is absolutely nothing to be gained by this, it is more sensible to heed the warning signals. Prolonged observation usually demonstrates how very tiring this can eventually become, in which case the 'vacant' eye will open spontaneously and the brain—by then accustomed to the monocular stimulus—will find it easier to adjust.

Far more useful than the teaching of this irrelevant and irksome ritual—which will come of its own accord readily enough—is to realise the importance of continuous adjustment of the controls during observation. Particularly under high power, the perpetual movement of the fine adjustment back and forth will enable a clearer picture of depth in the specimen to be built up. The to-and-fro movements of a mechanical stage will reveal a great deal of information about the histological layout of tissues or the orientation of a particle relative to others.

Slight changes in the adjustment of the iris diaphragm are often helpful, and even slight alterations in the focus of the substage condenser can sometimes reveal an unexpected increase in image quality. The basic settings for critical illumination must be borne in mind throughout, but the optimum setting for any control can only be realised by continual adjustments in either direction away from it. The car driver, after all, steers by perpetually moving the wheel back and forth, and his foot on the throttle moves it in and out, maintaining a position of optimum fuel supply.

Before looking at a specimen that has been set up by someone else, it is important to sit down (and not lean over); to look over the instrument to familiarise one's-self with the lenses in use and their adjustments; and then to position the hands so that the fine adjustment is comfortably held—and all this *before* bringing the eye to the eyepiece.

## THE BINOCULAR RESEARCH MICROSCOPE

The setting-up routine is carried out as before, but there are several adjustments of the binocular head that are necessary.

### *The interocular distance*

This is the control of the degree of separation of the eyepieces, which can be set to correspond to the distance between the eyes of the observer. The separation may be altered by rotating the eyepieces away from, or towards each other; it may be controlled by rotating a milled knob which moves the eyepieces together or apart on a screw-threaded shaft. The simplest form of adjustment has the eyepieces set in sliding holders that can be moved manually. It is important that the distance is carefully matched to the eyes so that visual overlap is assured. The brain can accommodate discrepancies of several millimetres, but unless the distance corresponds closely to the observer's interocular separation, eyestrain (in reality 'brain-strain') may set in.

## Compatible focussing

The paired eyepieces of a binocular microscope are generally fitted in a mounting that allows for one of them to be focussed by means of a screw-in mechanism. The correct means of utilising this facility is as follows. The microscope is focussed as described, with the eye adjacent to the focussing facility closed, or with a piece of opaque card held between eye and eyepiece, if closing it is inconvenient or tiring. One part of the object is carefully focussed—not a single cell, for instance, but a single cytoplasmic granule; not a given bacterium, but the optical section of a single spore, for example. Then the observing eye is closed, and the other opened. The eyepiece mount is screwed in or out to obtain exactly the same image focus. Both eyes can then be used in comfort.

But there is an important corollary to this. It is very easy, if no thought is given to the matter, for one eye to be in a different state of accommodation from the other and in this way a degree of incompatibility is introduced which can provide an additional source of strain. Therefore when making these adjustments, it is very important to relax the eye completely—ie to allow it to take its natural 'long-focus' position. For an eye in perfect condition this will be when it is focussed for infinity; though with advancing age or optical abnormality this is not practical. The effect is enough to allow the fully-relaxed eye to gain an impression of wide-angle viewing. When the eye is accommodating to such an extent that it is focussed for an object distance of, say, a matter of inches, not only is there an additional source of potential strain but the field diameter will appear to be small and compact. The universally-accepted notion of an 'image distance' of around ten inches in microscopy is pure invention—a convenience to aid calculations. The image distance will be anywhere within the normal focussing range of the observer's eye (between five inches and infinity, let us say). Indeed it is quite possible to go literally cross-eyed, focussing only a few centimetres from the end of one's nose, and still see a perfectly focussed image ...

Unless care is taken in focussing, then, it is a simple matter to set the instrument so that the two eyepieces require different degrees of accommodation from each eye. Alternatively, it is deceptively easy to fall into the trap of adjusting the image so that its distance from the eye is simply too small for comfort.

Given a setting that allows the eye to observe the image as though looking into infinity—staring at the night sky, for example—it is feasible to sit and carry out prolonged microscopical observations without significant visual fatigue. Microscopy can become less tiring than reading a book, in this way; and it is certain that a measure of the 'eye-strain' so commonly found amongst technicians and others who use the microscope is due to a failure to bear in mind the eye's accommodative capacity.

## *VIBRATION*

For research purposes the question of vibration becomes more important than it is in connection with the student microscope. The best way to test for the effect of vibration on a given model of microscope is to set it up on a firm base and then, whilst observing a specimen under high power, tap it lightly with the fingers. The aim is to impart a lateral movement at the level of the stage or focussing controls. The degree of vibration set up varies with the model, and this test—once the normal range has become familiar—should be part of the assessment of a new microscope that has been submitted for approval.

In practice the most frequent occurrence of vibration is due to the transmission of movements set up in the neighbourhood of the microscope bench. Passing traffic, heavy footsteps in a nearby corridor, apparatus elsewhere in the laboratory; these are the causes most usually found. The microscope bench itself must be heavy, and by its very inertia able to damp vibrations down. The resonance frequency of a bench made of lighter units may well help to propagate prolonged vibration. The microscope may be insulated from the floor by means of a bench mounted in rubber, on plastic foam, or even on inflated pads. A rubber-based block of concrete would give the most effective protection against vibration, but this is probably more than one would consider. Vibration hazard becomes a positive drawback in photomicrography, and it may even be induced by the operation of the camera or film-advance mechanism. It is useful to have the camera mounted separate from the microscope, with only a flexible (eg bellows) attachment between the eyepiece and the camera. In this way shutter vibration is minimised.

The methods by which commercial micrographic stands are constructed should be carefully assessed before purchase, and rigidity, freedom from resonance and tendencies to vibrate should be tested by the tapping method described above.

In addition, the shutter and mechanical movements of the camera should be carefully examined and tested. One very sensitive means of watching for vibration is to put a little mercury on to a watchglass with a large radius of curvature and sprinkle a few grains of chalk on to the surface. These can be illuminated laterally—a window will probably shed enough light—and even slight vibration will quickly show up as movements on the mercury surface. Alternatively, a circular cell (such as a washer, if there is no glass or plastic cell available) may be cemented to a slide and then filled with water until the meniscus is well above the edge of the drop. A few specks of dust can then be added in the same way, and vibration is quickly revealed when they are examined by transmitted light, using a middle or high-power objective. (Though any form of dust will suffice—and there may be enough on the water if it has been in contact with the atmosphere for any time—by far the most readily available is a little chalk brushed by light taps with the finger from a blackboard duster-pad). But when doing this test, remember that no microscope can be totally vibration-free. Every instrument will show its weaknesses if one of these methods is utilised, and the aim should be to obtain a comparative guide to the performance of different models, rather than trying to find a totally rigid stand. It is important to remember this, or you are likely to test model after model, and still fail to find one that seems satisfactory.

These general comments apply to the whole range of optical microscopes in use today. Perhaps the most exceptional is the Burch, which in operation is simple—but in terms of maintenance becomes exceedingly complex. Its two main components are the condenser and the objective mirror, both made of speculum metal and as we have seen, with a surface contour prepared with great care to within a fraction of one wavelength of visible light.

They are supported in heavy holders which serve to protect, to exclude dust, and to prevent extraneous light from creeping in and spoiling the image quality. These mirror units are mounted on the end of a steel cable which passes over a plastic pulley to a pair of lead counterweights which leave just a slight downward pressure on the focussing mechanism to eliminate backlash. A control knob to the left of the instrument, shown in the photograph, serves to raise and lower the whole assembly and the movement is transmitted through a train of pivots and bearings that are all fitted with steel balls. The main focussing controls are mounted at the other extremity of this extremely complex array, on the right of the device. A pair of control knobs —one each for the condenser and the objective—are clipped on to the fine adjustment heads so that they are easily accessible during normal operations.

The main focussing mechanism relies on a sliding wedge: as movement is imported laterally, a wedge that points 'upstream' to the movement is gradually raised and so transmits the movement—at right angles to the primary adjustment—upwards, in which direction it alters the relative positions of the two mirror units.

Each moving surface is fitted with the steel balls; so are the main load-bearing structures that carry the mirror units, and the holders themselves. Even the main pulleys bearing the wire cable from unit to counterbalance weight are mounted, not on some conventional axle or pivot, but on an annular ball-bearing mechanism that allows for slight lateral movements. And so, though the lenses in their housing weigh over 2.5kg, they move in their mountings with a precision and delicacy quite unlike anything else in microscopy.

Even here, in the largest and most cumbersome form of optical microscope ever made, the basic concepts apply. They apply too, to the many sophisticated forms of instrument available—such as the interference microscope referred to earlier. In the case of these types, manufacturers supply singularly detailed literature on the use and characteristics of each individual type and it must be emphasised that the instructions in the manual must be adhered to. Adjustments should be made according to the design specifications, and never from the instinctive dictates of an enthusiastic operator. Experience with one phase outfit, for instance, will supply general insight into operating principles—but quite different methods of adjustment and use may apply to the equipment of another manufacturer.

Only the handbook will reveal the correct procedures. If one is missing, the manufacturer will probably have a spare copy available. And a photocopy, mounted on card and covered with a plastic laminate, should always be available with the microscope itself.

## ADJUSTING THE ILLUMINATION

It is with the research microscope that the question of critical illumination becomes most important. Clearly, unless the light is adequately filling the objective aperture, there will be image degradation; and it is with the search for image quality that controversy has sprung up over the best methods of illumination. Reflected light is a less contentious issue, of course; the aim of gaining a suitable illuminated area is largely met by the choice of apparatus. But the use of transmitted light has been interpreted differently by two schools of thought. They involve interpretations that are more refined than the 'critical illumination' described on p140.

*Abbé-Nelson method*

This approach to the matter was originally devised as a response to the use of paraffin lamps which gave a uniformly bright, even, and flat source of light. *The light source itself is imaged in the specimen field.* Obviously this is excellent for light sources such as the frosted lamp mentioned earlier, or for lamps fitted with an opal or pearl diffusing filter. The greatest intensity of illumination is obtained by this means, though it introduces problems of its own. Chief amongst these are:

    flare

    uneven illumination of the field

    introduction of artefacts from the lamp image

    and the need to cut down the light transmitted by the condenser (by stopping down) leading to loss of resolution.

When the first gas mantles were introduced, a slightly defocussed condenser was used in order to avoid imaging the rectangular image of the mantle in the specimen plane, but when electrical lamps were introduced, the small area of intense brightness proved to be a considerable drawback. Abbé-Nelson illumination became obsolescent overnight—at least for routine microscopy.

*Köhler Illumination*

The presence of a condenser in front of an electric light source at once resolves some of these deficiencies. The light issuing from it—even if it has been produced by a very uneven source, such as a filament lamp—will be virtually uniform. The condenser, imaged in the specimen plane, will provide a circular disc of even light. In essence it becomes the light source, for all practical purposes. This fact had been realised before the turn of the nineteenth century, but it was the chief optical specialist of Messrs Zeiss who first published an account of the method in mathematical terms. He was Dr August Köhler, and the technique has been known by his name ever since.

In setting up the microscope by the Köhler method it is important to centre the whole system accurately. For this, a lamp diaphragm is useful, as it can be closed down to 'pinhole' dimensions and used as a guide-line in the adjustments. The lamp, condenser and diaphragm must be on a common axis, this must (after reflection by the substage mirror of the microscope) align exactly with the optical axis of the instrument. The method by which this end might be attained depends on the apparatus; the easiest way of centering the lamp is to remove the condenser and image the area of incandescence (through a neutral filter) in the microscope using a low power objective/eyepiece system. In Köhler illumination, *the lamp condenser is imaged in the specimen field.*

What of the merits of the two? It is reasonable to say, I think, that much of the controversy that has surrounded this debate has been pedantic and unnecessarily protracted. It is an essentially pragmatic matter, after all. The direct imaging of a very small area of incandescence in the specimen plane is ludicrous; and this invalidates Abbé-Nelson techniques for most modern applications. Similarly, the imaging of the lamp itself becomes impractical when a lamp condenser is in the way. The condenser *becomes* the light source. It is for this reason that the accurate aligning and adjustment of the lamp and microscope to provide usefully bright illumination for the purpose in hand has been referred to as 'critical illumination' in this book. Setting it up is a critical matter—but the haggling over details of a degree of refinement far removed from that which can materially alter resolution is to be deprecated. The sooner the Abbé-Nelson *v* Köhler controversy dies, the better.

### THE CONDENSER-LAMP DISTANCE

Any condenser of high quality has been manufactured to a design specification that had to take into account the condenser-lamp distance. It is important to make sure that this parameter is satisfied when assembling apparatus.

The distance varies, unfortunately, from manufacturer to manufacturer; and even from condenser to condenser. The United States producers seem to favour a 10in standard, whilst European standards range from 25cm and 30cm (Zeiss) to 44cm —measured from the stage, *via* the mirror—(Bausch & Lomb) and there are some others in between. Zeiss have provided a simple biconvex lens at $f=25$cm for their infinity-corrected condenser, the lamp housing being situated 25cm away, with this auxiliary lens in position. Other manufacturers using an infinity standard do not usually provide this lens. But the fitting of a spectacle lens of the desired focal length ($f=25$cm or 30cm) will enable these correction factors to be met if the lamp is the same distance away *measured along the optical axis, ie via the mirror*. Of course, if a parallel beam of light is emitted from the illuminator unit then this additional lens will be unnecessary.

The adherence to correction factors laid down by the manufacturer is obviously helpful in realising the best performance from the microscope; but time-consuming and intricate manipulations are usually tiresome in a routine laboratory. The basic scheme for obtaining critical illumination outlined on p140, modified if the apparatus demands it, is quite enough for bench use. It is only when using the most perfectly corrected optical equipment at its design limits that Köhler illumination need be accurately attained—and then it is of prime benefit when carrying out photography of the results. It is in this field—when the corrections for flatness of field are stretched to the limit—that care is most vital.

For general purposes the built-in illuminator is most handy, as we have seen. A useful alternative is the lamp unit that is built into a base on which any conventional microscope (ie with a U-shaped foot) can be mounted.

These designs have a built-in wheel of neutral filters (ground or opalised glass) or a selection that can be pivoted in or out of the visual axis before the light enters the condenser. It is useful to realise that the *clear* selection (ie with no filter in the light pathway) may give a beam that does not fill the full aperture of a good condenser. In this case, the ground glass screen(s) should be tried. The additional diffusion that this provides will enable the entire aperture to be adequately filled.

In conclusion, it is only by the use of separate lamp units—and not by an integral illuminator—that the full range of potentialities needed by the micrographer can be realised. This degree of flexibility is useful for precision microscopy of any kind.

But the built-in unit of the so-called 'photomicroscope' which is now so popular is highly successful in normal use. And the consistency of the results it can give far exceeds these standards customarily attained by busy microscopists with little time to spare.

CHAPTER SEVEN

# THE SPECIMEN

THE *raison d'etre* of microscopy is the examination of small structures. But few specimens are amenable to immediate scrutiny, as they are, unaltered; the vast majority are made as permanent preparations—microscope slides—and may be stored for reference purposes. A branch of technology (usually—if ambiguously—known as *microtechnique*) deals with the procedures involved. There are now many hundreds of them, perhaps many thousands if the various permutations are included, and countless reference books categorise them. We are concerned with the outlines of microtechnique in this book: and the standard procedures described later will suffice for routine purposes.

Leaving aside the complex literature that surrounds the permanent slide, there are many techniques and procedures that enable the microscope to reveal something of the nature of matter and life that are concerned with *real-time* examinations and analyses. Examining the fixed, sectioned and stained remains of a once-living tissue is a vitally important aspect of microscopy, to be sure; but the popular concept of research microscopy is too firmly embedded in this limited approach.

The purpose of microscopy should not be related so closely to the examination of permanent preparations. Rather, it should be aimed at the study of specimen material that has been suitably prepared for examination. We should be as interested in living microscopy as we are in the examination of permanent slides.

The microscope should be seen as a way of probing dust, fibres, and small structures from every facet of the environment. The more they remain free from artefact, the better.

The general standard of microscopy is not high. For all the legion comment over stains, mountants, illumination and the rest, there is precious little written about real-time, *immediate* microscopy; microscopy of the kind that is often supremely important. Drawings and diagrams of micro-organisms are almost traditionally inaccurate; the lay concept of micro-organisms and microstructure is generally lamentably inadequate; and even the most seasoned professional scientist is often able to misrepresent microscopic phenomena in a manner that seems elementary and embarrassing to any seasoned microscopist. If the camera of the telescope had suffered similar indignities it could be understood; but no, the corporate 'blind spot' is centered on microscopy. Writing elsewhere on the subject I was tempted to make the satirical suggestion that the unremitting misrepresentation of microscopic specimens was now part of scientific conformity; a ploy to confuse the outsider and preserve the exclusivity of the profession . . . yet sometimes that is almost how it seems.

A recently published schoolbook suggests that the microscope should be used to examine:

    a) a hair from your head
    b) a piece of white paper or blotting paper
    c) pencil or ink writing
    d) any other object

The last is a teaser: *any* other object? With a student's microscope of the kind featured in this unforgivable textbook almost everything around the schoolchild is incapable of being examined —desks, chairs, the floor, walls and ceiling; pictures, fittings, switches, cables and lamps; drawers and doorknobs. Even a smattering of microtechnique (the use of a razor to cut primitive sections, for example) would put the instrument into its true light.

As for the named specimens, they are exactly what the novice

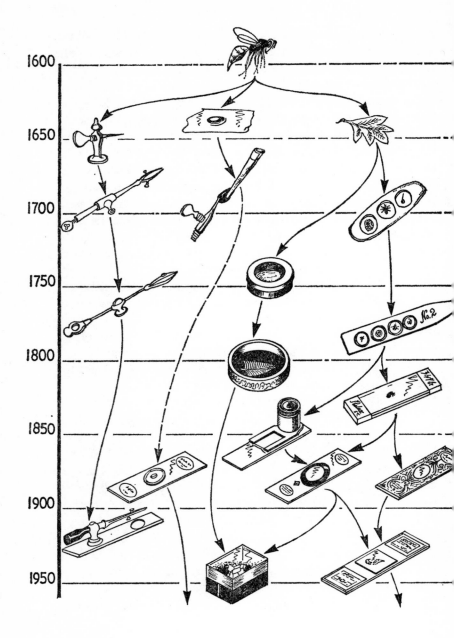

tends to look at. The torn edge of a piece of card can be used as a test object in some instances (see p138), but the surface of card or paper, along with the hair and the specimen of writing, are amongst the most disappointing of objects. Under low powers (and even under high magnification, to the uninitiated eye) they look exactly as they appear to the naked eye, but somewhat larger. No sudden detail is revealed, no microstructure is thrown up for the first time, no dramatic new insight is given. In this manner, the first glimpse of the world revealed by microscopy is a disappointment. It is exactly this trend that has, no doubt, given rise to the widespread lack of enthusiasm that is felt for the subject.

Finally, there is one vital factor that must be considered in any

Fig 9 *Evolution of the microscope preparation*—a diagrammatic survey of the main lines of development

*1600* The unaltered specimen. The first studies were carried out on objects (such as insects) magnified with a hand lens

*1650* By the middle of the century, a pin or stylus was commonly used to impale the specimen. Kircher examined fluid materials (containing nematodes etc), probably on mica slips, but most work was centered on the microscopy of detached, whole specimens such as leaves

*1700* The stylus was replaced by forceps around the turn of the century. This example was made in 1702 by Wilson. Leeuwenhoek pioneered the examination of fluid in capillaries, probably by dark-ground; they were mounted with sealing-wax on the stylus of his microscopes. Hartsoeker and others used hand-carved ivory slips, known as sliders

*1750* Forceps were produced for the compass microscope (as in this example by Adams, 1745). Small round boxes with mica lids appeared and became popular on the continent of Europe

*1800* The mounting of small, whole specimens in boxes reached its height. Cabinets to hold them were widespread in Europe. The ivory slider was occasionally replaced by glass slips cemented together by balsam

*1850* The 'compressorium' for live insects and other arthropod specimens became popular. Then the microscope slide and coverslip appeared. Sometimes these were used to make mounts of preparations in raised cells, but the routine mount was covered in patterned paper in simulation of the earlier slider mounts. The hanging-drop preparation, devised by Koch, became popular in the 1880s

*1900* Stage forceps were still in manufacture after the first world war, set on a perforated 3 x 1in base. The standard microscope slide with square labels at each end and a square coverslip became the norm

*1950* Microscope slides for routine purposes remained unchanged. Plastic boxes appeared for the mounting of mineral specimens, an up-dated form of the continental box mount

routine investigation. It concerns the 'first impression' of the microscopic image. Is it compatible with the normal object of its kind? Or is it in some way a departure from the norm? It is on this choice that the subjectivity of microscopical diagnosis depends, yet it implies something that is not normally taught. *How much variation within a given population of similar specimens is permissible before we begin to speak of 'abnormality'?*

For instance, in metallurgy the diagnosis of an alloy type depends on the relative size of grain formation (if present); the relative incidence of twinning; the presence or absence of coring ... yet it takes a very long while to recognise 'normality' or 'abnormality' in these contexts. Tissue sections in histology may show signs of pathological processes, yet the degree of variation within the norm can be considerable in some tissues, but negligible in others. Brain sections, for instance, can be said to be abnormal if even slightly different from a healthy specimen. But liver can show considerable variation in appearance in apparently healthy material.

In these somewhat extreme examples it is plainly apparent that the technician accustomed to examining brain material might tend to describe as 'abnormal' many sections of liver that were within limits; and conversely the specialist in liver histopathology might pass as normal many slides that his CNS colleague would successfully diagnose as diseased.

It is very important therefore to become acquainted with the degree of variation within 'normal' populations of differing types. The successful histologist does not—as generally happens in medical school—confine himself to a close scrutiny of a given slide. He looks in a detailed and conscientious manner at scores of slides of normal material, hunting about in them, following structures coherently, building up a spatial model in his mind of the relationships between tissue constituents, perpetually moving the slide and altering the focus. He examines slides from patients of different types, stained with different batches of stain so that he recognises over- and under-staining; he examines faded slides, remounted slides, dusty or scratched slides. He acquires insight into the various extraneous contaminants that occur, too. Pollen grains, surface vegetable hairs, wood dust, natural and artificial fibres are studied and understood. And in this way the picture of the norm is acquired.

*But more important to be learnt is the degree of variation within the norm.* The microscopist must assess the criteria by which he makes a judgement; and the criterion of conformity which applies so strongly to fern sporangia and *Pinus* pollen grains will be very much more rigid than those that apply to polished specimens of brass or tissue fibrocytes. This vital principle eventually becomes apparent, of course, in its own time; but how much more rapid is the growth of understanding in this field when that basic principle is borne in mind as an end to be consciously attained.

How does the formulation of a diagnosis arise in practice? Let us consider just one example of microscopic investigation as a case in point.

The material took the form of prepared vaginal smears that were brought to the author for examination. They apparently contained a puzzling feature—rounded foreign bodies were present in the smear, dotted about amongst the cellular and mucous material. Apparently similar bodies had been seen before. The suggestion was that they were some form of parasite, perhaps distorted by the artefact of preparation; but possibly something new.

The earlier investigations had taken the form of a detailed morphological study of the particles, including measurements of size, distribution and the surrounding cell population. That is a very sound, orthodox approach to the problem. But taking a fresh look at the specimen shows at once some key factors that point to the solution.

Immediate impressions of the slide were instinctively noted. It had traces of a mountant around the periphery of the coverslip. It peeled away easily and was probably DPX. There was a thumbprint on the coverslip, perhaps where it had been pressed down during mounting. This (since the consistency of the mountant showed that it had set sufficiently hard) was wiped away with a tissue, using a little mist from the breath to aid the process. The slide was examined with the high-power objective, then with oil. The cells were stained with haematoxylin and eosin, and the squames were somewhat keratinised for vaginal membrane, showing that this was in the last week or so of the menstrual cycle. Dotted about were the occasional rounded foreign objects.

The bodies were rounded, and showed tripartite markings. Three areas of flattening were visible, making one aspect into an obtuse pyramid.

This immediately suggested that the bodies had been produced in aggregates of four, the adjacent sides being formed each with three areas of flattening. Such a configuration suggests a meiotic reduction-division. Not only that, but the bodies were each possessed of a thickened cell wall which had stained well with the haematoxylin. A thickened wall suggests botanical origins.

These two observations lend themselves to one obvious interpretation: meiotically produced plant cells—pollen grains. Which raised problems of its own, for what were pollen grains doing in a vaginal smear? They were on closer examination seen to be indigenous parts of the smear, and had not simply floated in through a window on to the fresh smear. In addition the closer inspection confirmed that they certainly were pollen grains.

The appearance of fern spores was very similar to these bodies, and it happens that the related club-moss *Lycopodium* is the source of spores that are used commercially. A check with a reference slide from the reference collection confirmed the diagnosis: the bodies were no form of protozoan parasite, but were spores of *Lycopodium*. The uses of this powder are many, for the spheroidal shape of each particle gives it *en masse* a fluid and frictionless feeling, due to the ball-bearing action of each individual spore.

With this in mind it was possible, with a few telephone calls, to find that *Lycopodium* spores were used as the solid phase of a colloidal lubricant applied to condoms manufactured in this country. Comparison of the lubricant obtained from the manufacturers with the traces left in the vaginal smear provided confirmatory evidence. As it happens, the report in this case produced a hollow chuckle from the practitioner concerned, since the specimen had originated in a patient who (for reasons it would be invidious to enumerate) would not have been considered in the least likely to have had intercourse...

So this 'parasitic' inclusion turned out to have a very interesting story behind it. Not only that, but it leads to some interesting medico-legal consequences. For example, it has been held that the presence of recognisable human spermatozoa is the only evidence of sexual assault or rape that can be accepted in court as positive and unequivocal. But here we have an additional means of producing evidence in the event of a condom being used (and indeed the 'rubber fetish' might encourage certain individuals to prefer this in some instances of criminal rape). Not only this, but the origin of the condom might be identified too.

In further investigations it has been found that manufacturers use different solids in vaginal lubricants, including several varieties of starch as well as talc. Finally, what of the physiological effects of these agents on the vaginal mucosa?

So a single slide can produce a range of consequences. It is this application of microscopy as a discipline—and not merely as a handy technique—which can prove to be valuable in many fields of research.

## *EXAMINING THE SPECIMEN*

Though microscopes of the conventional magnification range (roughly x100 to x1000) are familiar enough, it is often in the range between that of a hand lens (x2 to x15) and that of the microscope that the real interest lies. Yet magnifications of, say, x20 are infrequently used.

There are several ways around the problem, the simplest of which is to acquire an extra long-focus objective for a conventional microscope stand. A lens rated at x2.7 or x3 is suitable for use with a x5 eyepiece and will provide an interesting extra facility.

Of course the most successful way of attaining it is to obtain a binocular, low-power microscope. This will give a wide field of relative flatness and the visual result is dramatically three-dimensional. Such microscopes are not necessarily expensive: though the superior research stand (with optics) will cost over £100, binocular microscopes for industrial and school use are available that cost—new—approximately one-tenth this figure. The corrections are good enough for visual use, but photographs are best not taken through this kind of optical system. The easiest way of taking macrographs, ie pictures (generally lit by diffuse reflected light) in the x5 to x25 range, is to utilise a projector lens as a single-lensed magnifier. Alternatively a short-focus lens may be used with a long bellows or extension tube on a conventional camera. With sufficient care, an enlarger may be used as a macrographic camera, with film or a plate in the position of the normal film holder and the specimen positioned on the baseplate.

It is in the low-power binocular microscope that specimens such as insects may be conveniently studied. They may be placed in a small plastic box as a means of restraining movement, or even held in the groove of a clothes-peg. Anaesthetising the insect may not be so successful, and ether will very likely kill it outright. But the most useful way of slowing down insect movements is to rely on their body temperature variation and to cool them by leaving them in a refrigerator for an hour or so.

The low-power microscope is of widespread applicability in fields as diverse as forensic microscopy, textiles and printing, palaeontology and so forth. Indeed, on acquiring one it is most instructive to re-examine familiar specimens as permanent preparations (organ sections, plant specimens etc) since the lower magnification reveals many structural details that a more conventional approach precludes.

## THE SOLID SPECIMEN

Some specimens, such as minerals, may be examined by the microscope as they are. The only problem that is likely is the difficulty of getting a large specimen on to the microscope stage, still having enough room to focus. The low-power microscope (*vide supra*) is often mounted on an arm, or fitted with a ring-foot which has a large, unobstructed central area and so it can be easily brought over the specimen. Occasionally, with a conventional microscope, it is feasible to put the specimen underneath the stage and bring it into focus through the condenser aperture in the centre. Otherwise, if this is possible, a small fragment may be detached for examination.

Metal materials are generally polished and etched before being examined. A suitable specimen is removed from the sample and cut so that it fits comfortably on the stage. It is then ground to a flat surface and subsequently polished to give a mirror surface. It helps if this is exactly parallel to the opposite face, so that the light from a metallurgical objective is reflected back along the optical axis, giving a spectacularly illuminated specimen.

The initial grinding and flattening can be done by abrasives such as 'wet and dry' paper of successively finer grades, but the final polish is given by the use of a wet polishing cloth mounted on a horizontal rotating disc that is electrically powered. The abrasives of choice for this final polishing stage are:

    (i) rouge
    (ii) levigated aluminia

The surface can, in this way, be rendered so smooth that scratches are not visible even at x1000 magnification. But it appears predominately featureless when freshly prepared and, though particulate inclusions or cavities are visible easily enough, etching of the surface is necessary to reveal microstructure such as grain boundaries. Suitable etchants vary with the metal under examination. For steel, nital or alcoholic picric acid are among those that are suitable. Nonferrous metals may demand alcoholic ferric chloride, or ammoniacal solutions of (a) hydrogen peroxide or (b) ammonium persulphate. Alexander's reagent, containing acetic and nitric acids, is also widely used. Any acidic liquid may have uses as an etchant in metallurgy, of course.

| NAME OF ETCHANT | COMPONENTS | AMOUNTS USED |
|---|---|---|
| Nital | nitric acid in alcohol | 3% $HNO_3$ |
| picric | picric acid in alcohol | varies |
| Alexander's reagent | acetic acid (75% glacial) | 30ml |
|  | nitric acid (bench conc.) | 20ml |
|  | acetone | 30ml |
| ferric chloride | ferric chloride | 5g |
|  | ethyl alcohol, (ethanol) to make *slightly acidify with HCl* | 100ml |
| peroxide | 0.880 ammonia solution (aq) | 25ml |
|  | water | 75ml |
|  | $H_2O_2$ ('20 volume' soln aq) | 0.1ml |
| persulphate | 0.880 ammonia solution (aq) | 10ml |
|  | ammonium persulphate | 10g |
|  | water                 to make | 100ml |

## *THE PARTICULATE SPECIMEN*

Very small dust particles as smoke can be demonstrated by means of reflected light. The original form, dubbed an '*ultramicroscope*', was devised in 1903 by Zsigmondy, and took the form of a small chamber filled with smoke, which was illuminated by a lateral beam of light. This was brought to a fine focus by means of a lens, forming a bright area on which the objective was focussed. Small, brightly illuminated particles of smoke may be seen moving rapidly about in a state of Brownian vibration. It has been calculated that particles only 40 Å in mean diameter have been visualised by this method.

The effect, named after an early investigator, is known as Tyndall scattering. It is most pronounced for particles of the same order of magnitude as the wavelength of the illuminating beam, though light may be scattered at comparatively high rates by particles—as we have seen—very much smaller than that.

It is important to realise that at this order of size, the dimensions of the particles as visualised are not a function of the size of the object particles themselves, but vary with the numerical aperture of the lens system. The brightness of the particle varies with a factor of its dimensions relative to the illumination wavelength.

The brighter the illuminant, the better the chances of seeing signs of very small particles. The smallest of all have been observed with the aid of an arc-light struck between carbon arcs; but sunlight would probably be even better. For this purpose it would best be captured by a mirror driven on an azimuth telescope mounting, in which form its movements would exactly cancel out those of the sun as viewed from the earth's surface. The device is known as a *heliostat*.

At such orders of object size it is naturally impossible to speak of 'resolution'. The reflective particle is not seen by transmitted illumination, as it is too small to cause any interruption in the advancing wavefront. The focussed image of the light rays reflected from it is *not* an image of the particle itself, therefore. It is an *antipoint*, and confers no direct information about the size of the particle. Nor does it reveal anything of the particle size though periodic changes of brightness could possibly be related to the real shape of an elongated or unidirectionally-reflective particle rotating as it moved. That is only an academic question, however.

The ultramicroscope gives a fairly direct confirmation of the existence of atoms and molecules; it is certainly the easiest and most convincing demonstration of molecular bombardment due to heat energy. And it has provided the nearest to direct evidence in support of Avogadro's hypothesis. In 1811 an Italian theoretician named Avogadro recognised that the easiest way to explain direct proportionality of volume in gas reactions, was to assume that equal volumes of gases contain an identical number of molecules under identical physical conditions. The number of molecules in one mole of a substance (ie its gram-molecular weight) is known as *Avogadro's Constant*.

An ultramicroscope of the form we have described is rarely used in modern times. It is not much more than an interesting demonstration device, really, and is not a serious research tool.

Far more widely used (though still something of a cinderella) is the ultramicroscopy of particles in a liquid, instead of a gaseous, medium. This necessitates the use of a dark-ground condenser, a technique described in the following paragraphs.

*The dark-ground condenser*

The term 'dark-field' has often been applied in this context but as we shall see, it would seem to be incorrect. It is the ground—ie the background—that is dark. The field, or at least those parts that are reflective, is bright; in some cases very much so. For this reason the term 'dark-ground' seems to be the only correct one, and is used throughout this book.

The equipment for dark-ground microscopy as described on p116 is capable of revealing much information about ultramicroscopic structures in liquid suspensions. If a multipurpose condenser is fitted, then it must be centered and adjusted according to the maker's handbook. There are no standard procedures for these very varied devices.

The setting up of a dark-ground condenser is a task that must be done correctly if the best results are to be obtained. Many of the comments listed below will apply equally to the fitting and adjustment of a multipurpose condenser, of course.

(1) The substage is lowered and swung out, away from the microscope. The condenser is removed and placed to one side. For safety it may be replaced in its holder, if one has been fitted to the interior of the microscope case, or alternatively it can go into the box from which the dark-ground condenser has just been removed for use.

(2) The dark-ground condenser is substituted. Note: there are

many diameters of condenser on the market; standardisation here has not yet been widely attempted. It is often very difficult to fit one manufacturer's condenser into the substage of a microscope made by someone else. In an emergency, it may even be possible to stand the dark-ground condenser *on the upper surface* of the substage condenser mount and use it in this way—if it doesn't fit in the ordinary manner. This is at best a clumsy alternative.

(3) The condenser is raised until level with the stage. A low-power objective is then engaged in the microscope nosepiece, and this is focussed carefully on the upper surface of the condenser. In the centre of the upper element is a small, engraved circle. Using this as a guide, the condenser is accurately centered. In this operation, it is assumed that the oil-immersion objective shares a common axis with the objective now in use, or the centering will be approximate only.

(4) A drop of oil is placed on the condenser, and the slide added to it. The substage mirror is adjusted to throw a strong beam of light through the condenser. It is visible as an area of light in the preparation film: it is very likely annular in appearance.

(5) The condenser is now adjusted to the position where the lighted area on the slide is a small point of illumination, and not a disc or an annulus. *This may be impossible.* If it is, a thinner slide must be substituted.

(6) The oil-immersion objective is then engaged. This must be:
  i. an objective specifically designed for the condenser; the aperture may approach 1.20 in some models;
  ii. or an objective in which has been fitted a conical stop. Most objectives rated as oil-immersion lenses unscrew, allowing this stop to be inserted. It effectively reduces the aperture to 1.0 or thereabouts.

  Note: If several are available, it is always possible to gently ream metal away from the inner surface of the aperture, testing as you do so, until the dark-ground condition is *just* lost. A thin coat of black matt paint, by closing the aperture very slightly, will restore the dark-ground appearance to the image and may allow a figure in excess of NA=1.0 with certain condensers.

iii. an objective fitted with an iris diaphragm may be used instead. This rarely-used device allows the stop to be adjusted until the maximum aperture compatible with dark-ground microscopy is obtained.

(7) The coverslip is oiled, a single drop of immersion oil being gently dropped directly on to the brightly illuminated spot, and the dark-ground objective is lowered into it.

(8) Focussing is then carried out as for light-ground microscopy, described earlier.

It is obvious that a very high-intensity light source should be used. If a conventional lamp is too dim, it is worth trying a projector. There is likely to be one in the laboratory somewhere, and a 1000 watt projector can become a useful high-intensity microscope lamp without modification, though it may be better in some models to remove the focussing lens of the device. Sunlight has been used with success for dark-ground microscopy, though it is best to filter it through a copper sulphate solution (p111) to remove infra-red radiation. If the illumination under oil is disappointing, this could be due to the lack of a common optical axis. If a non-centerable nosepiece is fitted (eg a rotating triple nosepiece) the mirror must be slightly adjusted until maximum brilliance is obtained. Dark-ground condensers are not fitted with a stop or iris diaphragm: the object of the exercise is to get as much light as possible on to the specimen particles, and since the background is always black there is no virtue in fitting such a device.

*Slides for dark-ground microscopy*
Slides must be quite clean and free from particulate blemishes or scratches that, by reflecting light of their own, can cause interference with the image quality. Methods of cleaning slides are given on p119. A drop of the suspension is taken with a pasteur pipette and placed in the centre of a clean 3 x 1in slide. A No $1\frac{1}{2}$

coverslip is taken from its container and carefully polished with a non-linting cloth, the use of mist from the breath being used if necessary. It is then stood on the slide adjacent to the drop, and gently lowered with a dissecting needle on to the drop, which spreads by capillarity.

*Care is essential*: if the drop is too large (and it will be, if a full-size drop from a pasteur pipette is used) the coverslip will float and move during adjustments. If this happens, the slightest change in focussing control will cause the particles in the field of view to move at high speed and it may take time for the field to settle down again.

On the other hand, if the drop is too small (and this can happen if too much liquid is withdrawn into the pipette) then capillarity will not spread it to the edges of the slip, and evaporation losses from the edge that the fluid *has* reached will cause the preparation to lose volume and eventually to dry up altogether.

If drop control from a pasteur pipette proves to be difficult, a pipette drawn out finely and nipped off where it fits in the penultimate small hole of a standard Imperial drill gauge—this will give it the property of delivering a drop approximately 0.02ml in volume—may be easier to use. Such a pipette, the use and preparation of which is described in standard bacteriological texts, is known as a 'fifty dropper' since it delivers one-fiftieth of a millilitre. For normal use a x15 eyepiece will be found to give an image that is somewhat faint. A x8 or x10 is preferable.

The distribution of particles such as latex, Indian ink, or the oil droplets in milk, may be very successfully demonstrated by this means. The use of phase microscopy, which is very popular for such specimens, shows them as such indistinct particles surrounded by prominent haloes that the observations are greatly hampered. And of course this is incapable of showing particulate matter of submicroscopic dimensions. Dark-ground microscopy as a demonstration of Brownian movement is particularly interesting.

## BIOLOGICAL MATERIAL—LIVING

Far too few biology microscopists examine living material. It is essential to become conversant with the living cell as an entity, yet our standard teaching of histology, cytology—even protozoology and the study of other single-celled organisms—is hidebound by a reliance on permanent preparations. The number of schoolchildren, technicians and doctors who have looked at blood smears is incalculably vast. But the number who have ever studied living blood cells by phase, interference, or dark-ground microscopy seems to be vanishingly small. The most frequent candidate for living microscopy is probably *Hydra viridis* which has been obtained from a biological supply firm. *Hydra* is notoriously susceptible to fumes such as those of acetic acid and a high percentage of those that are examined by students are actually dead, or moribund at least.

For teaching purposes there are many easily-available living specimens. Some of them, and the most suitable means of observing, are listed below.

| SPECIMEN | SOURCE | MICROSCOPY |
|---|---|---|
| Bacteria | Teeth-scrapings | Dark-ground |
| ,, | Decaying matter in pond | ,, ,, |
| Protozoa | Surface pond scum | Dark-ground, phase, Rheinberg, etc |
| Algae | ,, ,, ,, | ,, ,, ,, |
| ,, | Filamentous bloom in fresh water | Light-ground |
| Blood cells | Man (from fingerprick) Insect (from haematocoel) Worm (dissected) | Dark-ground, phase, Rheinberg, etc |

The most widely thought-of specimens of living human cells are squames from the cheek lining. But, like all easily-removed mucosal cells, they are metabolically inert, virtually dormant or dead, and give no idea at all of the nature of a living cell. The haematocytes of insects under dark-ground microscopy are extremely interesting, as they are active at normal temperatures and do not lose activity as human blood cells tend to do, when they cool down. Though we have used locusts from a laboratory culture for this purpose in the past, any insect will suffice.

The stagnant water at the bottom of a pond or a ditch—or even gathered from a moist gutter—will often provide masses of micro-organisms for examination. Alternatively, infusions made by boiling hay or other dry vegetable matter (this being the method favoured by pioneer protozoologists—the organisms were known as 'infusoria' for many years) may be left until organisms become manifest.

Active organisms may be slowed down effectively by adding some dilute agar to the preparation, or by adding strands of cotton wool or filamentous algae to the mountant water.

For the examination of living blood cells there are some interesting possibilities. One departure is to dilute the blood sample in a pasteur pipette with several times its own volume of physiological saline solution. It can then be examined by interference microscopy, phase or Nomarski contrast. A personal opinion, since we have used it very often for this purpose, is that darkground microscopy is by far the most dramatic and revealing of all. Some very beautiful studies of blood are possible using this technique.

*An original technique for single drops of blood in conditions of dynamism* has proved to provide a new facility for research. Originally it was evolved for the purpose of providing a moving sample of blood, using only a single drop. In this way we can do away with ring-circuits etc which are so cumbersome. An outline of the technique is given here; further details are published elsewhere (see bibliography):

(1) A sample of blood is obtained by finger-prick from the nail bed (in man) or from any other suitable source.
(2) Cleaned 3 x 1in slides are used to pick up a drop of blood. They are set down on the bench for a period that has to be found empirically, and which varies with environmental factors such as temperature and humidity etc, but will be

somewhat in excess of 5 minutes.
(3) A coverslip is added to the drying sample. The slide is then placed under the microscope for examination.
(4) Observations using the high-power phase objective are carried out as blood, squeezed by capillary attraction and the weight of the coverslip, breaks through the peripheral region of gelled blood and flows through making undulating pseudo-capillaries as it does so.

In this way the dynamic conditions pertaining to blood cells during the coagulation process may be studied. This is a uniquely interesting technique, mainly because it gives—for the first time—a repeatable means of making haemodynamic studies with the microscope that is rapid, foolproof and requires no extraneous apparatus; and secondly because only a small drop of blood is needed for the test, thus removing the need for venepuncture. Since evolving this method we have obtained some very graphic pictures of blood as it coagulates; and the use of dark-ground microscopy—as described in the previous section—enables one to observe the previously undetected microfibrils which attach the erythrocytes to the fibrinous reticulum. (See also p91).

*Hanging-drop mounts* are very practical for the examination of small samples of motile organisms from culture, and may be used for blood (diluted with physiological saline) if stationary cells are to be observed. The technique was devised by Koch, but is a direct descendant of Leeuwenhoek's primitive methods. It is based on the examination of a single drop of fluid on the glass surface. But a simple, uncovered preparation has several disadvantages:

(i) evaporation losses can easily occur
(ii) the lens may mist because of water condensation
(iii) the objective may come into contact with the culture organisms—and they may be pathogens
(iv) dust may be introduced, and oil immersion objectives are clearly unusable

Note: Water-immersion lenses, which can be immersed in the aqueous mountant containing the organisms, have been produced. For microscopists they are curiosities, nothing more.

The hanging-drop technique is carried out as follows:

(1) A slide is prepared with a cell cemented to its upper surface. If thin cells are used, a slide with a concave depression may be selected.

(2) A round coverslip, selected to fit the cell, is set down on the bench and a drop of the specimen material is placed in the centre of it. The drop must not be too large.

(3) The rim of the cell is lightly smeared with Vaseline and the slide is then inverted and lightly pressed on to the coverslip. The whole mount can then be set the right way up again, and the slide is ready for examination.

Oil immersion can be used with success, but because it is somewhat pointless to oil a condenser to the slide (because of the air space above) high-resolution work is limited, and dark-ground virtually impossible. The observation of organisms at the very edge of the mountant-droplet may supply examples that, due to the thinness of the film, are relatively stationary. In them, motility mechanisms may be studied with ease.

*Microscopy of living tissues in vivo*

Extensions of this approach to living microscopy are many. The blood flow in the capillaries of the ear of rabbits and other experimental animals may be studied by shaving the skin and confining the meatus in some form of restraining and supporting chamber—a technique described at length in specialist literature —but there are very instructive alternatives available.

The oldest (since it dates back to Leeuwenhoek's day, and was popular from then until the 1920s) was the study of capillary flow in the fins and tail of a fish. For ease of manipulation I think it is hard to beat the tail of a tadpole, when available; high-power studies—even using ordinary high-power transmission microscopy—are dramatic and instructive. More complex methods involve the study of blood flow in the mesentery of experimental animals, a difficult and tedious process in many ways, but this technique has uses in certain research programmes. It is quite superfluous to teaching and demonstration, however, when much more readily available material is at hand.

*Living tissue in vitro*
Tissues may be cultured in small chambers that are obtained for the purpose, and examined directly by the microscope. A warm stage heated to 37° is necessary for human or other mammalian tissues. This technique is most applicable for time-lapse cinemicrography and some very dramatic views of cell division have been taken in this way. The easiest method is to culture cells in Hank's solution, or some form of suitable growth medium, in a culture chamber that contains coverslips. These can then be removed, one side cleaned, and the culture mounted face down in the culture liquid. Plant tissues may be cultured from meristematic tissue explants. In most cases inorganic solutions are used, perhaps with the addition of hormones such as 2,4-dichlorophenoxyacetic acid. The details of tissue culture are not of immediate interest to the microscopist, and we will not attempt to examine the subject here. But detailed monographic studies of plant and animal techniques are readily available.

*(Supra) vital staining*
One of the simplest means of staining living cells is to feed particles of carmine to *Paramoecium* or some other ciliate. The pathways followed by food vacuoles can in this way be studied clearly. And there can be no doubt that vital staining is a technique capable of providing much interesting information, as well as being a means of obtaining very striking preparations. Yet the subject is—to borrow a term from another book!—unfashionistic. Microscopists are not in the main aware of the potentialities of the technique; it is rarely taught, infrequently referred to in textbooks. Yet its eclipse is, one hopes, a temporary affair.

Slides for this purpose have to be cleaned. Two main *methods of cleansing 3 × 1in slides for vital staining* have been recommended:

(1) Stand the slides for five days in chromic acid, after which drain them, and rinse thoroughly under the hot water tap. Stand overnight in hot water which is allowed to become cool (it should be distilled or deionised water at this stage) and then store in strong (say, 90%) alcohol. When required they are picked up by forceps, wiped with a cloth and flamed.

(2) A quicker method is to wash slides with soap solution and hot water, rinse, and dry in a clean cloth. They are then rinsed in two changes of xylene. This does not serve to rid them completely of traces of alkali or grease, but the pH changes are of little consequence for a short-term fluid mount and the grease that may remain is in the form of a monomolecular trace that does not greatly interfere with the spreading of the liquid then applied.

The slides are then *treated with stains* ready for use. The most widely used stains are neutral red and Janus green. Concentrations of the stains in solution vary with stain and also with the manufacturer's recommendation. Instructions should be carefully followed, since though a slightly understained specimen will merely take longer to develop colour in the cells, an overstained slide will kill the cells rapidly and invalidate the results.

Stock solutions of 0.4% Janus green and 0.25% neutral red are prepared and, for use, these are added to alcohol, in the proportions given:

| | | |
|---|---|---|
| 0.4% Janus green | — | 0.07ml |
| 0.25% neutral red | — | 1.75ml |
| absolute ethanol | — | 10.00ml |

The slides are treated by placing a drop of the stain solution at one end, and drawing it along so that it is spread in the manner of a drop of blood for a blood smear (*vide infra*). The alcohol evaporates, and the stain-coated slides may be stored for prolonged periods in a dust-free container.

For use, a drop of the cell suspension is placed on the stained surface of the slide and a cover slip is added. It is advisable, in cases where evaporation might cause hypertonicity, to seal the edges of the preparation with wax or melted Vaseline. After 20 minutes or so the cellular inclusions should be stained. During this time the preparations should be kept at temperatures normal for the cells and out of light.

The technique is of use in differential diagnosis of blood samples, in which case a fat-free sample of blood is obtained by lifting the finger-prick sample from the skin, having cleansed the area with an ether-soaked pad. The results are gratifying. Cytoplasmic granules within the polymorphonuclear granulocytes show clearly; yellow in the neutrophil, bright orange in the eosinophil, maroon in the basiphil. Mitochondria show clearly as ellipsoidal or elongated blueish structures in the lymphocytes, and the reddish vacuolar structures and blue mitochondria of the monocytes are also clearly visible.

## *Material from agar culture*

Apart from culture on the slide itself, many organisms are cultured on agar, including fungi, bacteria, algae, and germinating spores from green plants. For many applications, a *temporary mount* is very practical. Some of the germinating spore material, for example, is removed with a platinum loop and placed in a drop of water on a slide. A coverslip is added, and the specimen examined in the normal manner. Oil immersion should *not* be used; the aqueous mountant prevents high resolution and introduces chromatic effects, and the coverslip will tend to drift due to the higher viscosity of immersion oil.

*Agar mounts* are an entirely new departure and were developed in my laboratory for certain observations on radiating yeast and fungus colonies. Many of the spoliage organisms found in foodstuffs containing starch (eg sausages) are yeast-like species that grow in nutrient agar as penetrating, stellate colonies. It is instructive to examine the relationship between the cells as their radiating branches grow and expand through the growth medium. However it is clearly impossible to remove them *en masse*, as they are deeply penetrating the agar medium. So smears or any other form of permanent preparation are impossible.

Microscopy of the colonies *in situ* is generally impossible, too. Since the branches penetrate deeply, a high-power objective will be unable to focus on them without becoming embedded in the medium itself. Oil-immersion is ruled out, and the slanting orientation of the colony gives a very limited depth of field. The thickness of the agar medium causes a severe loss in image quality, due to lenses corrected for higher refractive indices.

The agar mount if carried out carefully can give very pleasing temporary preparations and these can be made permanent if the need arises. The stages in making the preparation are outlined below:

(1) On a clean 3 x 1in slide, a small amount of standard crystal violet is deposited. The amount of stain and its concentration are not critical. I tend to dip a clean dissecting needle into some normal (bacteriological 1%) crystal violet and then tap it on the slide; this leaves a small amount behind which is adequate for the technique.

(2) The platinum loop is flamed and, whilst still hot, it is used to scoop out a small area of agar with the desired peripheral area of the colony in it.

(3) This is then transferred to the slide, adjacent to the stain, so that the stain and the agar containing the colony branch are in contact.

(4) A coverslip is added, leaning across the agar specimen.

(5) A cover (such as a metal jar cap) is stood on the slide, in order to protect the coverslip, and the mount is placed in the autoclave in a strictly horizontal position.

(6) After being autoclaved for 5-10 minutes at any standard pressure (since they are all enough to bring the agar gel, above 100°, at which point it will be molten), the pressure is lowered gently to ambient, and the slide removed.

(7) No further manipulations are attempted until the slide has cooled down, ie when the agar is once more solidified.

This technique gives well-stained cell bodies in an orientation close to that pertaining in the culture. For phase or interference microscopy, the stain may be omitted, as the contrast can be increased by other means; but the cytological details are clearly shown by the crystal violet and stages of division are easily observed. In an emergency, it is always possible to place the slide at stage (4), above, across the corner of a tripod, and then warm it cautiously with a very small blue bunsen flame, bringing the flame under the slide only intermittently, until the agar melts. At this stage it may well boil, too; but if the slide is cooled by blowing on it the mount may settle down well enough to give fairly clear examples.

This rapid method is not as good as the other; but of course it is impossible to obtain slides of such colonies by any previously described method and, for preliminary use, it may prove valuable.

## BIOLOGICAL MATERIAL—DEAD (TEMPORARY MOUNTS)

Many biological specimens may be subjected to microscopy whilst in some state of chemical alteration, staining or other process that has in some way altered their condition from the living state, but which does not amount to the making of a permanent preparation. This is an unusual form of microscopy these days. Material may be examined alive; or it may be examined dead, in the form of permanent mounts. Tests of material in the field of microchemistry can be used to identify specific compounds or classes of compound, some of which can be mounted temporarily. Some forms of freezing-microtomy give material suitable for temporary mounts, and routine plant sections may be mounted in glycerine for preliminary examinations. We have already considered agar mounts (*supra*).

*Plant sections*

Sections of plant tissues may be cut by hand, using a well-stropped 'cut-throat' razor or a new safety razor blade. The blade, the material, and the sections are all kept wet with water throughout. The scheme for staining and mounting such temporary sections is as follows:

(1) Cut sections and transfer them with a fine paint-brush to a watchglass of distilled water.

(2) Select sections in which all cell outlines are clearly visible. Transfer them to stain solutions as follows:

| SOLUTION | TIME | MOUNT |
| --- | --- | --- |
| Lugol's iodine (stains cell structures brown, starch navy blue) | 5min | glycerine* |
| Aniline sulphate (stains lignin yellow) | 5min | glycerine* |
| Phloroglucin (stains lignin violet) | 5min | transfer to fresh watch glass containing few drops HCl; then mount in glycerine* |
| Schultze's solution (stains cellulose and starch blue) | 5min | glycerine* |
| Sudan IV (fats, oils fatty acids orange) | Soak in 70% alcohol for 10 seconds then stain in solution for 15-20min | glycerine* |

*Pure glycerine may be used; but a 50% glycerine solution aqueous may cause less distortion.

(3) Sections are lifted from the stain with a dissecting needle with as little stain as possible; they may be rinsed in 50% glycerine aqueous if too much remains. A round coverslip is most suitable for the spreading drop of mountant.

(4) Surplus mountant is removed with blotting paper if necessary to prevent the coverslip from floating. The slide is then labelled with a grease-pencil (or adhesive label).

*Dry material*

Temporary mounts of dry material, such as spores, fibre, sporangia, dusts, mineral fragments etc, may be made by clearing the material in xylene for a few moments and then mounting it in immersion oil. Hollow specimens such as fern spore-cases are best placed in a closed bijou bottle or other screwtopped jar, and then warmed in the incubator overnight to cause bubbles to escape from within.

## BIOLOGICAL MATERIAL—DEAD (PERMANENT PREPARATIONS)

*Negative stains*

The simplest form of stain is probably the negative staining technique applied to motile bacteria. A drop of bacterial suspension is mixed on the slide with Indian ink or some similarly opaque liquid. The drop is then allowed to dry and the thin black film that remains is interrupted by clear outlines wherever bacteria are present. They are seen as clearly as stars in the sky, as punched-out silhouettes; flagella may be seen with clarity. This is not a widely used technique in today's laboratories, but yields some graphic results if carefully employed.

(1) Side by side on a clean 3 x 1in slide place a small drop of bacterial suspension and another of (a) Indian ink or (b) 10% nigrosin solution.
(2) Mix the two thoroughly with the platinum loop.
(3) Allow to dry (a process that occurs rapidly).
(4) Examine with oil immersion microscopy. It is not necessary to mount the preparation, the oil may be applied to the slide direct; but for best results it is as well to make a permanent preparation by mounting in Canada balsam or DPX.

*Simple stains*

The simple stain is invariably applied to bacteria obtained from pure culture, where morphological studies only are required.

(1) In a small drop (a loop-full) of distilled water is placed a very small inoculum of the culture. Enough to adhere to one edge of the loop is ample. The suspension is then mixed and well spread over an area about as big as the coverslip.

(2) The film is left to dry. This happens more slowly than in the procedure for negative staining (above).

(3) The slide is passed through the bunsen flame several times, until the film of condensate re-evaporates, and the bacteria are fixed.

(4) The stain is added *only* when the slide has cooled. Examples are:

> fuchsin
> methylene blue
> safranin
> neutral red
> malachite green
> methyl violet

The stain is left for 1 minute.

(5) The surplus stain is well washed off under the tap, and the film is cautiously blotted by being laid, face-downwards, on clean filter-paper.

(6) The slide is observed direct, or mounted in Canada balsam or DPX as described in the previous section.

A clear preparation showing the bacteria—associated structures, such as capsules or spores, may also be visible—is the result. Note that methylene blue and methyl violet tend to fade in the acid mountant Canada balsam after a few months.

## Fontana's method for spirochaetes

Films or smears containing spirochaetes may be stained by a silver precipitate that shows up the organisms as brown, undulating spirals. They do not stain easily by any conventional means, although the genus *Borrelia* will take carbol fuchsin or the conventional Romanowsky stains. Fontana's method is a complex procedure, but can give very good results.

(1) Fix the fresh smear in a fixative solution containing 2% formaldehyde + 1% acetic acid (aqueous), and change the fixative by flooding the slide with fresh solution every 60 seconds for 3 minutes.

(2) Wash off the fixative with a liberal stream of alcohol. Leave some alcohol on the slide for 3 minutes, then pour off the excess and remove the rest by flaming the slide.

(3) Add mordant solution, 5% tannic acid + 1% phenol (aqueous).

(4) Support the slide in the corner of a tripod (p180) and gently heat the slide until the mordant steams. Leave it like this for 30 seconds.

(5) Wash with distilled water from the wash bottle and allow to dry in the vertical position.

(6) Flood the slide with silver nitrate. The solution is made as follows:
    (i) prepare a 0.5% solution of $AgNO_3$ aqueous.
    (ii) add to it a 10% ammonia solution until the precipitate that forms does *not* redissolve when a few drops of ammonia are added.

(7) Heat this slide, as before, until the steam rises and leave it in this state for 30 seconds, heating gently from time to time.

(8) Rinse gently under the tap for several minutes, blot carefully and allow to dry.

(9) The slide may be mounted in Canada balsam or DPX as before.

## *Differential staining of films*

A simple distinction between two groups of bacteria was proposed by Christian Gram, a German microbiologist who published his technique for the first time in 1884. He was searching for a method of preparing double-stained preparations of pyogenic cocci that matched those of Koch and Ehrlich in their work with the tubercle organism. Gram went on to find that his primary stain for the bacteria (gentian violet mordanted with iodine in potassium iodide solution) was retained only by some species; others readily lost the stain when the smear was rinsed in alcohol. It is now known that the difference is dependent on the presence of magnesium ribonucleate, which stains deep violet-black with the primary stain. Organisms such as *Staphylococcus, Bacillus* and the yeasts come into this category and are said to be gram-positive.

Those that do not contain magnesium ribonucleate lose the stain, and are counterstained with a red dye (usually safranin or neutral red, though a diluted form of carbol-fuschsin may be used). Organisms such as *Neisseria, Proteus, Salmonella* and the coliforms come into this gram-negative category—*and so do normally gram-positive organisms that, through age or an artificial medium, may lose their ribonucleate content.* The technique that Gram first published has been variously modified; a personal version of it is given below.

(1) Heat-fixed films (see p183) are flooded with 0.5% crystal violet for 1 minute.
(2) The stain is poured off, and then washed with Lugol's iodine (or Gram's iodine, which is 3 times more dilute).
(3) Stain with the iodine solution for 1 minute.
(4) Wash with alcohol until stain ceases to come away, or in acetone.
*Caution*: prolonged washing will tend to remove the stain from all organisms; if a sudden surge of the violet colour does not come away almost at once, the organisms are gram-positive. Acetone is particularly rapid in its action.
(5) Wash under the tap.
(6) Stain with safranin or neutral red for 1 minute.
(7) Wash under the tap, blot, dry and mount in DPX.

Gram-positive organisms stain very deeply by this method, and appear purple, or near-black. Gram-negative organisms are red. Nuclei are red (roughly the same colour as the gram-negative organisms) and cytoplasm is pink. An alternative counterstain is eosin, which may be available in a routine histology laboratory.

Differentiation of *Mycobacterium spp* is based on the organisms' great powers of stain retention when treated with low-pH solvents. They are, due to the waxy coat (which also explains much of their resistance to drying etc), also very difficult to stain. In the early days of bacteriology, the tuberculosis organism was notoriously difficult to identify because of this.

The staining technique relies on the use of carbol-fuchsin solution. This stain is made by dissolving 1% basic fuchsin and 5% phenol in a 10% solution of alcohol in water.

(1) The dried film need not be fixed.
(2) Carbol-fuchsin (above) is applied to the slide and it is then kept steaming (p184) for 5-10 minutes.
(3) The slide is then washed under the tap and decolorised as below:
    (i) 20% aqueous $H_2SO_4$ will show all acid-fast organisms except *M.leprae*;
    (ii) 3% HCl in 95% alcohol (aqueous) will show *M.tuberculosis*;
    (iii) 5% $H_2SO_4$ will show *M.leprae*.
(4) The specimen is counterstained with methylene blue or malachite green. Mount as usual.

In this way, acid-fast organisms stain bright red; other structures (cells, non-acid-fast bacteria, etc) stain blue or green. It is possible to stain sections with haematoxylin to demonstrate cell/organism relationships.

*M.tuberculosis* (which is still frequently—though wrongly—designated by the abbreviation '*Myco*') may be present in very small numbers, and has recently been found to stain with auramine O. It fluoresces bright yellow under the ultra-violet lamp, and may become much easier to locate.

*Protozoa as films*
The most revealing means of observing protozoat is by the use of interference-microscopy on living specimens. The preparation of permanent mounts is a branch of microtechnique that involves

some intricate processes, as the cells are delicate and are easily disrupted. For routine microscopy, it is possible to obtain very satisfactory slides by the new method described below:

(1) A drop of water containing the organisms (eg *Vorticella* or monads) is spread on a slide and allowed to dry, forming a conventional film.

(2) The slide is fixed with an aqueous fixative such as formal-acetic (Fontana) or formal-acetic-alcohol (Ford). The fixative is left on for 2 minutes.

(3) Wash the film cautiously in distilled water, dripped from a washing bottle, so that organisms are not dislodged.

(4) Flood the film with:
    (i) crystal violet (dilute the bench stain with twice its volume of distilled water), or
    (ii) Ehrlich's haematoxylin.
Leave the stain on for (i) one minute or (ii) five minutes.

(5) Wash cautiously with tapwater. Crystal violet may be washed in alcohol or acid-alcohol, if the film is overstained. Erhlich's haematoxylin may be removed with acid-alcohol in the event of overstaining, the film being subsequently washed for 5-10 minutes in alkaline tapwater.

(6) Counterstain with eosin for 1 minute.

(7) Dry *without blotting*. Mount in DPX.

Nuclear structures are well revealed by this method, but care must be taken to avoid overstaining.

The same technique has also been used in studies of small algal cells.

*Blood smears*

A method has been widely popularised in which a small finger-prick specimen of blood is placed between two coverslips. As they are slid apart, a smear of blood cells is left on each. I still prefer to use the traditional two-slide method, ensuring that the slide edge does not crush its way through the cells in the process. This is probably a somewhat pedantic refinement. The smear, according to this method, is made as follows:

(1) Cleanse the nail-bed of the middle finger with ether.

(2) Using a Hagedorn needle, a fresh glass splinter, or a Hemolet® lancet, a sharp stab wound is made through the skin.

(3) Drops of blood are picked up near one end of 3 x 1in slides without touching the skin with the slide.

(4) A second slide is placed across the slide bearing the blood drop, and 'backed up' towards the drop until contact is made, when the blood drop spreads along the contact edge (see figure).

(5) This spreader slide is moved rapidly and firmly along to the other end of the specimen slide, thus leaving an even film of cells behind. The smear dries almost at once.

(6) Fix in *either* methanol, *or* undiluted stain (see table) which —since it is dissolved in methanol—acts in the same way, for 2-3 minutes.

(7) Stain according to the following table:

| STAIN: | TYPE: | DILUTION: | TIME: | WASH: |
| --- | --- | --- | --- | --- |
| Giemsa | Romanowsky | 1:2 ($H_2O$)* | 5min | distilled water until pink |
| Leishman | ,, | 1:2 ($H_2O$)* | 4-12min | ,, |
| Wright | eosin + methylene azure | 1:1 buffer pH 6.4 | 5min | tapwater 2min |
| Jenner | compound | 1:2 ($H_2O$) | 3min | distilled water until pink |
| Jenner-Giemsa | mixed | stain as above with Jenner's solution (diluted) for 1min, followed by Giemsa's stain (diluted) for 15min. Wash, blot and dry | | |

* Colour changes in cells may become apparent in certain areas, due to tapwater solutes. A buffer at pH 6.8 should be used instead.

(8) When dry, mount in Euparal or DPX. Preparations may be examined direct, using immersion oil applied to the slide.

These methods are also applicable to blood smears containing parasites (eg *Trypanosoma*).

## Tissue smears

Smear techniques are used for several tissues with advantage. Isolated neurones may be seen in smear preparations of central nervous tissue, and the method is also of use in preparing slides of spermatogenisis in the earthworm. Crushed anthers will give good material for floral chromosome studies, and dried preparations of metaphase plates are used as a standard method in human chromosome counts. There are many methods of staining in the literature, ranging from simple stains (such as methylene blue, particularly applicable to the Nissl granules of neurones) and reagents including osmic acid ($OsO_4$) which react with fats and oils, to conventional double stains, *qv*. Orcein is used for plant chromosomes and for zoological material too; but crystal violet may give clearer results in smears of animal cells containing metaphase plates. We have used crystal violet for human chromosome-count material and the results are clearer than acetic orcein.

## Tissue sections

*(a) plant sections cut by hand.* Many plant tissues, due to the rigidity of the cell walls in botanical specimens, may be cut by hand into 70% alcohol as described on p181. The razor is honed on a whetstone to give a fine edge, and then stropped on leather 20-30 times. Delicate plant tissues, such as leaf material, may be supported in a piece of firmer material such as elder pith or carrot.

The sections are then transferred to staining solutions with a seeker or a camel-hair brush and are dehydrated in alcohol or cellosolve. A basic scheme is as follows.

(1) Cut fine sections into a petri dish of 5% alcohol (which acts as fixative as well as merely supporting the sections).
(2) Transfer likely-looking examples to a watchglass of 70% alcohol and inspect under low-power microscope to select the best sections.
(3) Stain with a single example listed in the table below:

| STAIN | SOLVENT | AFFINITY |
|---|---|---|
| Aniline blue | alcohol | sieve plates |
| Bismarck brown | alcohol | cell structures (nuclei and walls) |
| Eosin | water/alcohol | cell walls |
| Haemalum | water | nuclei (blue) |
| Haematoxylin | alcohol | nuclei (blue) |
| Light green | alcohol or oil of cloves | cell walls |
| Methylene blue | water/alcohol | nuclei |
| Safranin | water/alcohol | lignin, suberin, nuclei |

(4) Rinse in 90% or absolute alcohol, then
  (i) differentiate in acid alcohol if overstained, or
  (ii) proceed direct to (5).
(5) Dehydrate in absolute alcohol (two changes).
(6) Clear in xylene.
(7) Mount in DPX, Canada balsam or Euparal.

*Dehydration with cellosolve.* This material is ethylene glycol monoethyl ether and it is miscible with alcohol, oil of cloves and xylene in all proportions. Plant sections may be transferred direct to cellosolve after staining and can be mounted in DPX without clearing. However it is best to dehydrate in cellosolve and then clear in xylene in the normal way.

*Counterstaining in oil of cloves.* Oil of cloves is a gentle (ie non-distorting) clearing agent miscible with alcohol and xylene. It is a solvent for light green, and for this reason it is possible to counterstain in the following way:
(1) Take a section stained with safranin, haemalum, haematoxylin (or aniline blue) to absolute alcohol.
(2) Transfer to a solution of light green in oil of cloves. Stain for 2-5 minutes.
(3) Transfer to oil of cloves to remove excess stain and clear.
(4) Mount in Canada balsam or DPX

(*b*) *Plant sections by microtome.* Plant material can be sectioned best by a microtome. In this case it is best to embed the material. Many polymer materials have been marketed for this purpose, but the traditional embedding agent remains paraffin wax. This is used most widely in histology laboratories and remains the agent of choice. Embedding methods are discussed in the following section, p193.

The plant tissue is either cut in a conventional mechanical microtome as animal material also is, or it may be sectioned in a hand microtome. This device does not give very satisfactory results with animal tissues. It consists of a specimen carrier which may be extruded gradually by means of a calibrated screw. This is turned after each section has been cut, raising the specimen slightly above the cutting plate. Hand microtomes of this sort became popular at the turn of the last century and are principally used in schools and college laboratories.

(c) *Zoological material as sections.* Animal specimens are sectioned on a mechanical microtome. The most popular model is the Cambridge rocking microtome, though there are several other types available including sledge and swinging-arm designs. Tissues are fixed (as appropriate) and may be stained in bulk:

(1) Fix in a bichromate fixative, or similar, for several days, depending on the size of the tissue; it should not be more than 1-2cm across.
(2) Wash in running tap water for 24hrs.
(3) Transfer to 25% alcohol for 24hrs.
(4) Transfer to 50% alcohol for 24hrs.
(5) Transfer to the solution below for 14 days (approx):
Scott's haematoxylin, 1vol + acetic acid (aqu) 2%, 5vols.
(6) Wash in running tap water for 24hrs.
(7) Transfer to eosin (0.5% or 1% aqu) for 24hrs.
(8) Transfer to alcohols of the following strengths:
25%, 50%, 75%, 90% absolute,
each change containing 0.5% eosin. This prevents loss of the counterstain by solution. Leave for approx 12hrs in each.
(9) Transfer to absolute alcohol for a final 6hrs.
(10) Clear in xylene for 24hrs (either use a good volume of the agent or use two changes).

(11) Embed in wax as described below.

Sections may then be cut and mounted direct on to slides. After drying they may be steeped in xylene or benzene to remove the wax, and then mounted direct in DPX. Such a method is rarely used but is of great value in the preparation of serial sections, sections of embryos, or slides for use in quantity for commercial or class distribution.

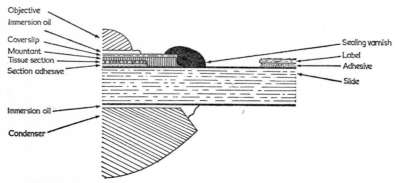

Fig 10   Transverse section of a typical slide in use (not to scale). The varnish ring is usually absent in modern preparations

*Routine zoological sections*

There are so many staining methods available for use in connection with human or animal tissue sections that it would be impossible even to summarise them; but without question the most widespread technique that the microscopist will have to hand is double staining with haematoxylin and eosin. Haematoxylin solutions are varied. Some stain progressively, whilst others are regressive. Again, some have to be well 'matured' before they are ready for use, whilst iron haematoxylin has to be made immediately before it is to be employed. And—though all stains should be filtered—volatile alcohol in some solutions evaporates so rapidly as to make this a chancy business, and decanting has to be relied on instead. For routine use, Delafield's or Ehrlich's haematoxylin are popular, though Scott's formula is sometimes

recommended. Basic staining procedure is as follows:
(1) Take the section through alcohol to 25%.
(2) Stain in haematoxylin for 5 minutes.
(3) Rinse and blue in tapwater or water from a wash-bottle containing a drop or two of ammonia.
(4) Differentiate, if necessary, in acid alcohol.
(5) Counterstain with eosin for 1 minute.
(6) Dehydrate successively, clear in xylene, mount in DPX or Canada balsam.

*Injection of animal material*
Organs or entire small specimens are kept warm after death by immersion in a saline water-bath. A medium is prepared—ready-made as a rule—containing gelatine and carmine as a colouring agent. This is liquid at 37.°

A wide-bore needle is used to inject the medium into a prominent afferent blood vessel and the specimen is subsequently fixed and hardened in 90% alcohol (cold).

*Thick* sections are cut with the microtome. It is advisable to check the thickness as cutting proceeds by laying one fresh example on a slide and treating it with xylene. The blood vessels show up in the form of a red network, and if sections are thin (2-5 µm) then the continuity will be lost. Thickness is not a pronounced limiting factor in this method, as the bulk of the tissue remains unstained and therefore does not impede the passage of light rays through the matrix. Some very graphic studies of lung, kidney and liver are possible in this way. The technique is generally ignored today, which is unfortunate.

*Embedding*
Tissues for embedding in wax are treated with a fixative selected for the particular stain envisaged. Routine fixative solutions may be selected from those in the laboratory.

After fixation and washing, the tissue is taken through the alcohols as described on p191 to xylene. It then goes into a concentrated solution of the wax in xylene for 24 hours. Following this, it is transferred to three changes of wax kept molten in a hot chamber at around 60°C.

The last stage is the pouring of molten wax, containing the specimen, into a mould. This is best made from two L-shaped metal elements standing on a glass surface. The mould may be dampened with a detergent solution to aid release of the wax when solid. The rectangular block containing the tissue specimen is plunged into cold water for an hour or two.

Sections, when cut in a room roughly of 20° temperature, will form into ribbons that can be collected on glazed paper and are easily mounted on albumen-treated slides. A trace of dilute albumen solution dried on a slide will greatly aid the adhesion of the section.

There are many choices of mountant available. Aqueous mountants are of two types: those made from gelatine, and those that are gums. The gelatine-containing mountants are melted by the application of heat, usually by standing the bottle containing the mountant in a beaker of water at around 50°C. The liquid mountant can then be removed on the glass rod, in the normal manner, and a small drop applied to the slide. After adding the specimen and adjusting its position, a coverslip is gently lowered on to the drop, which spreads rapidly. Within ten minutes, at room temperature, the mount has solidified, and may be sealed with slide varnish (nail-varnish is a useful substitute) or examined direct. Such mountants may contain anti-fungus materials—such as thymol—or a stain that is sufficiently dilute to remain near colourless in the drop, but which will colour the specimens adequately. This is a rarely-used technique today.

The gum-based mountants are used in a similar way, though there is no need for them to be warmed. They set by the gradual evaporation of water and form polymers that are not, as a rule, very near glass in refractive index. For specimens that cannot be dehydrated they may be useful, though they are inapplicable for routine use.

Liquid mountants, such as lacto-phenol, are applied to the slide in small amounts. In this way, when the slide is sealed, there are only slight risks of moving the coverslip and spreading the mountant. If too much is applied, the surplus being removed with a cloth or pipette, the slide will retain traces of it and any sealing varnish will refuse to adhere firmly to the glass. Such preparations leak badly.

Euparal is a resinous mountant that is miscible with alcohol as well as with xylene. Its refractive index is low for critical work, and there is nothing to commend it for many routine uses (for example, in the mounting of blood smears). But in specific applications it is valuable, particularly where xylene-clearing is to be avoided.

Canada balsam remains a popular mountant, even though it was regarded as 'imperfect and awaiting substitution' a century ago! Surplus mountant may be chipped off with a stylus or scalpel, the remaining traces being removed with a trace of xylene and a cloth. After mounting, it is best to place the slide in an incubator or oven—but care is necessary, for both xylene and benzene (an alternative solvent for balsam) are highly inflammable. It is traditional to stand slides on the warm region of a microscope lamp and allow it to harden off in that way—but that is an unpredictable way of doing things.

The most popular mountants these days are those consisting of a solution of a plastic (such as polystyrene) in xylene—DPX is an example. The initials here stand for *Distrene* (a proprietory polystyrene): *Plasticiser: Xylene*. Mountants of this kind have the advantage that they can be peeled off when dry, so that the removal of surplus traces outside the coverslip area is facilitated. These mountants are more liable to form 'strings' when handled; they lack the 'dropping' characteristics of balsam. However they are neutral, and therefore do not lead to such a degree of fading as balsam.

Certain proprietory mountants have been developed (such as MAC by Messrs G. T. Gurr) that do not require a coverslip. In this way a drop is placed over the specimen, and the preparation is put away in a dust-free cupboard to dry. The drop flattens and leaves a circular area similar to a normal mount. In spite of its advantages for quick, routine work, this method has remained unpopular. A similar principle is found in the aerosol mountant Acrylek launched by Fisons late in 1972. Coats of the mountant are sprayed on to slides in batches, and no coverslip is added.

Mounting has moved into a more futuristic era with the development of methods that eliminate the need for a coverslip. However, the traditional approach remains the most widespread and for day-to-day microscopy it is likely to remain so.

Microscopy remains the developing, flexible, accommodating discipline it always has been. Modifications to procedures are always possible, and microscopists quickly learn short cuts and private processes—some of them quite unmentionably hair-raising when you consider the problems that might arise. Particular care is always needed when pathological material is being handled. Not only are the results of any microscopical examination important, in that they dictate the decisions made by the physician in respect of the patient, but the very materials themselves are—by definition—capable of causing disease in certain circumstances. All suspect waste materials should be autoclaved for 20 minutes, or steeped in chromic acid or a strong lysol solution for at least 24 hours. Used slides with tubercle on them are best destroyed in the incinerator, if there is plenty of burning material already in it to prevent them from slipping out at the bottom. It is high time there were legal restraints placed on the conduct of research in microbiological laboratories and it is still regrettably true that there are episodes of infection resulting from the deficient techniques used in many establishments.

Microscopy calls for care, diligence, and open-mindedness, but the applications are infinite, and the future will undoubtedly give as much opportunity as the past for original research and discovery.

# SELECTED BIBLIOGRAPHY

## HISTORICAL WORKS

Bradbury, S. *The Evolution of the Microscope* (1967). Oxford
Clay, R. S. & Court, T. H. *The History of the Microscope* (1932)
Dallinger, W. H. (ed) & Carpenter, W. B. *The Microscope and Its Revelations* (1901)
Dobell, C. *Antony van Leeuwenhoek and His Little Animals* (1932)
Doetsch, R. N. *Microbiology: Historical Contributions 1776-1908* (1960). Rutgers, US
Ford, B. J. 'Medical Microscopy Yesterday and Today', *Medical News* (August 1966)
  'The March of Science', *History of the English-speaking Peoples* (1971). 2592-2595.
  'Recreating the Pioneer Microscopists' View', *New Scientist* (1971). 51 (763) 324-325
  'A Reconstruction of the Microscopic View of Nature Two-and-half Centuries Ago', *British Journal of Photography* (1971). 118 (5793) 682-685
  *The Lens Reveals* (1973)
Hooke, R. *Micrographia* (1961). New York
Martin, J. H. *Microscopic Objects Figured and Described* (1870)
Rousseau, P. *Histoire de la Science* (1945). Paris.
Wood, J. G. *Common Objects of the Microscope* (1889)

## CONTEMPORARY MICROSCOPY

Allen, R. M. *Photomicrography* (1958). New York
Bennett, A. H., Jupnik, H., Osterberg, H. & Richards, O. W. *Phase Microscopy* (1951). New York
Biological Stain Commission. *Biological Stains* (1961). Baltimore
Cameron, E. N. *Ore Microscopy* (1961). New York
Cosslett, V. E. *Modern Microscopy* (1966)
Ford, B. J. *Microbiology and Food* (1970)
Kerr, P. F. *Optical Mineralogy* (1959). New York
Loveland, R. P. *Photomicrography, a Comprehensive Treatise* (1970). New York
McCrone, W., Draftz, R. G. & Delly, J. G. *The Particle Atlas* and *The Particle Analyst* (1967). Michigan
Price, G. R. & Schwartz, S. 'Fluorescence Microscopy' in *Physical Techniques in Biological Research*, ed Oster & Pollister (1956). New York
Waksman, S. A. *Soil Microbiology* (1952). New York
Wilson, G. S. & Miles, A. A. *Topley and Wilson's Principles of Bacteriology* (1957). Baltimore
Zobell, C. E. *Marine Microbiology* (1946). Waltham

*See also Journals including :*

The Microscope
Journal of the Royal Microscopical Society (*now* The Journal of Microscopy)
Journal of the Queckett Microscopical Club
Journal of Applied Bacteriology
Proceedings of the Royal Microscopical Society

# INDEX

## NAME INDEX

Abbé, 53ff
Adams, 42, 43
Aidie, 44
Amici, 54, 82
Argand, 69
Avogadro, 168

Bauer, 45
Bausch, 94
Beeldsnyer, 49
Bell, 63
Bichat, 44
Bohmer, 76
Bonanni, 53
Brown, 50
Burch, 82, *103*

Chester-Beatty, 83
Cock, 17
Cohen, 74
Corti, 74
Cuff, 42, *51*
Culpepper, *38*

van Deijl, 49
Delafield, 192
Delft, 26
Descartes, 9, 11, 68
Dobell, 27
Dyson, 90

Ehrenberg, 73
Ehrlich, 184

Fox-Talbot, *85*

Franchot, 69
Fraunhofer, 45

Gerlach, 74
Giemsa, 188
Göppert, 74
Goring, 44
de Graaf, 30
Gräbe, 75
Gram, 184
Gravesend, 27, 29
Grew, 25
Grubler, 76

Hank, 176
Hartig, 74
Hartnack, 54
Hartsoeker, 37
Hill, 73
Holland, 29ff
Hooke, 11, 16, *34*, 68
Huygens, 69

Jenner, 188
Judson, 76

Kepler, 9
Koch, 184
Köhler, 95

Lealand, 47
Leeuwenhoek, 26ff, *33*, 68
Leishman, 188
Lieberkühn, 68
Liebermann, 75

Lister, 44ff
Lomb, 94
Lugol, 181

Malpighi, 23, 25
Martin, 42ff, 161
Marzoli, 49
McArthur, 80
Muller, 59
Musschenbroek, 36

Newton, 123
Nobert, 45ff, 54
Nomarski, 90

d'Orleans, 77

Pasteur, 77
Pepys, 17
Perkin, 75
Pluta, 90
du Pont, 9
Powell, 46
Power, 17
Pumphrey, 66

Reade, 43
Reeves, 17
Rheinberg, 116
Rochester, 27, 29
Roe, 83
Romanowsky, 188
Ross, 46, 54

Scarlett, *38*
Schott, 57

Schultze, 181
Schwarzschild, 82
Scott, 191
Smith, 46
Sorby, 58
Stephenson, 56
Swammerdam, 24, 25, 68

Talbot, *85*
Tinney, *51*
Tolles, 57, 121
Töpel, 121
Tortona, 37
Tulley, 49
Turrell, 79
Tyndall, 166

Varley, 47
Vickers, 80

Wainwright, 71
Whitbread, 77
Wenham, 54
Wheeler, 62
Willcocks, 82
Wilson, 37, *52*
Wollaston, 45, *89*
Wratten, 71
Wright, 188

Young, 45
Ypelaar, 60

Zeiss, 55
Zernicke, 84
Zsigmondy, 166

## *SUBJECT INDEX*

aberration, chromatic, *50*
  spherical, 50
achromatism, 49, *50*
acid, acetic, 64
  chromic, 177
  hydrochloric, 64, 67
  nitric, 66, 67
  picric, 111
  sulphuric, 66

  tannic, 184
acridine orange, 96
adjustment, coarse, 132
  fine, 133
agar mount, 178
algae, 20, 32, 64, 65, 172
alum, 64
aluminia, 165
ammonia, 74

# INDEX

analyzer, *89*
*Anguillula*, 65
aniline, 75
   sulphate, 181
*Anthophysa*, 31
antipoint, 167
*Apis*, 24
aspherizing, 83

bacteria, 59, *86*, 172
balsam, *85*; *see also* Canada
barrel, 16
beam, zero-order, 87
bee, 24, 30
Bell's cement, 63
bell-glass, 63
Berlin black, 62
binocular eyepiece, 77
Bismarck brown, 190
blood cells, 172, 178
   flow, 40
   smears, 187ff
   vessels, 73
brass, 28
brine, 20
butterfly, scales of, 44

camel's hair, 66
camphor fluid, 63
Canada balsam, 57, 60, 182ff
candles, 70
capillary flow, *33*, 175, 176
carmine, 65, 73, 193
castor-oil, 65
cellosolve, 190
cells, 59; *see also separate entries*, eg blood
cellular structure, 12
cement, Bell's, 63
chromosomes, 189
circlip, 40
*Cladophora*, 32
clearing, 61
cloth, 28ff, *34*
condenser, 39, 52, 78, 114, *192*
   achromatic, 115
   aplanatic, 115
   dark-ground, 115, 168ff
   oil-immersion, 117
confervoid, 64

contrast, aplitude, 90
   bright, 88
   dark, 87
   Nomarski, 90
cork, 12ff, 31
copper sulphate, 110
cover-slip, 53, 59, 120ff, *192*
   round, 181
   square, *159*
critical illumination, *see* illumination
crystal violet, 179

dark-field, 168; *see also* dark-ground
dark-ground, 72, 90
daylight, 19
dehydration, 61
desmids, 65
diatom, 55, 65
dichromate, 111
differential, *see* stains differential
diffraction, 32
DPX, 182ff
draw-tube, 48
dye, 73-6
dyes, Judson's, 76

entozoa, 65
embedding, 61, 191-2
eosin, 185
epithelial cell, *86*, *91*, *92*
erythrocyte, 45, 91
etchants, 166
ether, 60
ethylene glycol mono-ethyl ether, *see* cellosolve
euchrysine, 96
euparal, 188ff
eyepiece, 130-2
   compensating, 127, 131
   flat-field, 131
   Huygenian, 131
   special, 135
eye-strain, 148
eye, injury to, 72

fabrics, 29, *34*
ferns, 65
fibre, 182
filter, 71
   coloured, liquids for, 111

# INDEX

gelatine, 107
glass, 107
  infra-red absorbing, 111
  Rheinberg, 116
  transluscent, 19
  ultra-violet absorbing, 111
fish, *33*, 175
flagella, 182
flea, 24, 43, *86*
flea-glass, 36
fluorite, 58
fluorochrome, 76, 95
fly, 9, 21, 44
foot, 132
forensic science, 96, 164
fuchsin, 76, 183

gasteropoda, 66
generation, spontaneous, 18
glass, flea, 36
  flint, 57
  soda, 57
  special, 57
globulist theory, 35
glycerine, 63, 181
grass, 30
gum, 61
gutta-percha, 67

haematin, 76
haematoxylin, 76, 187ff
*Haematoxylon*, 76
hair, 35, 40
Hank's solution, 176
heliostat, 117, 167
histology, 13, 182ff
holomicrography, 90

illumination, types of, 102ff
  Abbé, 152
  adjustment of, 151
  critical, 140, 151
  Köhler, 152

illuminator, 114
illusions, optical, 108
immersion, 54, 56
Indian ink, 182
indigo, 73
insufion, 35

infusoria, 59, 173
injection, 25
insect, 23, 61
interferometer, 90
iodine, 181
iris diaphragm, 37, *78*, 141ff
ivory, 59

Janus green, 177
*Juncus*, 40

kidney, 23

lamp, arc, 69ff, 106
  Argand, 69
  blue light, 96
  coiled-coil, 72, 105
  coiled filament, 72, 105
  discharge tube, 73
  duplex, 69
  electric, 70
  filament, 71
  Franchot, 69
  gas, 70
  grid, 71
  mercury arc, 107
  methylated spirit, 70
  oil, 20, *52*, 68
  paraffin, *52*, 69
  quartz-halogen, 73, 105
  ribbon filament, 105
  solid source, 105
  triplex, 69
  tungsten arc, 107
  ultra-violet, 95, *103*
  xenon arc, 107
  zirconium, 107
laser, 90
lead, 28
leaf, 66
lens, achromatic, 43
  apochromatic, 58
  arsenic, 16
  bull's-eye, 71
  cracked, 139
  flask, 19
  fluorite, 58
  grinding, 30
  gum, 16
  negative, *14*, 43

# INDEX

objective, *89*, 124ff, 192
  oil, 16
  plano-convex, 16, 19
  positive, *14*, 43
  resin, 16
  salt, 16
  single, 26, 28
  water, 16
  water-immersion, 175
lichens, 66
lighting, incident, *85*
light, green, 190
  reflected, 21
  ultra-violet, 83; *see also* lamp
locust, 173
louse, 24, 30, 40, 43
lysol, 194

malachite green, 183ff
Malpighian corpuscles, layer, 23
mauve, 75
mesentery, blood in, 176
methyl violet, 183ff
methylene blue, 111, 183ff
mica, 21, 40
*Micrographia*, 12, 19, 26, 28, *34*
microscope, achromatic, 46
  automated, *104*
  Bausch and Lomb system, 94
  binocular, 77, 146
  blue light, 96
  body, 128
  Burch, 72, *103*
  case, 140
  compass, 69
  compound, *78*, 100
  Culpepper-Scarlett, 38
  dark-ground, 90
  double, *51*
  drum, 41
  fluorescence, 76, 95
  Hooke's, *34*
  infra-red, 94
  interference, 88, *89*
  low-power binocular, 164
  lucernal, 43
  McArthur, 80
  misuse of, 144
  new universal single, 42
  opake solar, 43

  pocket, 81
  Powell & Lealand, 52
  reflecting, 54, 81, *103*
  research, *104*, 146, 151
  scanning, *104*
  screw-barrel, 37, *52*, *86*
  second-hand, 138
  simple, *33*, 98
  solar, 43
  split-beam, *89*
  ultra-, 91, 166
  ultra-violet, 94
  Wilson, 99
  zoom, 135
Microscopical Society, 47
microscopy, fluorescence, 95
  teaching of, 97
microtechnique, 156
microtome, 190-1
mirror, 19, 68, 112
moss, 9, 11, 65
mount, agar, 178
  liquid, 31
  techniques of, 194-5
  temporary, 180ff
mountant, 18, 183-4, 194
*Mucor*, 17
muscovy glass (muscovite), 21

naphthalene monobromide, 58
Newton's rings, 123
neutral red, 177, 183
nigrosin, 182
nipple, 71
nose-piece, 42
numerical aperture, 55-6

objective, 53, *89*, 124-8
  achromatic, 124
  apochromatic, 127
  flat-field, 125
  fluorite, 127
  oil-immersion, 125
  plano, 125
  special, 128
oil, castor, 65
  cedar-wood, 56, 121
  immersion, 121
  synthetic, 122
paper, oiled or waxed, 19

parfocality, 142
patch-stop, 91, 116
pathological material, dangers of, 194
*Pediculus* (*see under* louse)
penderocyte, 91
petal, 66
petri dish, 189ff
pewter, 28
phase contrast, 84-8
  negative, 87
  positive, 87
phloroglucin, 181
photomicrographs, 43
plant sections, *see* sections
pointolite, 71
polarizer, 89
pores, 12, 30
potassa, 62
preparation, hanging-drop, *158*
  permanent, 23, 163ff, 194
protozoa, 35, 74, 172

quinine bisulphate, 111

rape, 163
razor, 12, 20, 21, 67
red blood cell, *see* erythrocyte
refraction, *10*, 13, *14*, 15
refractive index, 56
*Rhizobium*, 17
rotatoria, 67
rotifers, 35
rouge, 165
Royal Microscopical Society, 47
Royal Society, 11, 30

safranin, 183ff
*Sambucus*, 40
scattering, Tyndall, 166
screw-barrel, 37, *52*, *86*
sealing wax, 15
section, 12, 189-92
sections, plant, *85*, 180-2
seminaria, 18
shellac, 63
silk, 75
slide, 60, 86, 118ff, 170, 192ff
slide, evolution of, *158*
slider, 39, *51*, 59, 61, 99

smears, 189
snowflakes, 21
sodium iodide, 123
  nitrite, 111
solution, *Schulzez's*, 181; *see also*
  stains
specimens, dried, 18
  solid, 165
spermatozoa, 35, 163
spicules, *58*
spider, 30
*Spirogyra*, 32, 35
spontaneous generation, 18
sporangium, 182
spores, 18, 182
squames, *86*, 172
stage, 135
  mechanical, 134
  sliding, 79
stain, differential, 184-6
  Gram's, 185
  negative, 182
  simple, 182
  supra-vital, 176
stands, 132
stratum spinosum, 23
stylus, 11
substage, 48, 53, 69, 78, 134, 168-70
Sudan IV, 181
sulphuric acid, 66, 181
sunlight, 19, 68

tannic acid, 184
tartrazine, 112
telescope, 9, 44
temporary mounts, 180
test plate, 45ff, 54
textile, 28, *34*
thymol, 112
time-lapse, 176
tripoli, 15
tube length, 122
turpentine, 62

varnish (rings of), 60
Venetian glass, 13
vibration, detection of, **148**
voltage regulator, 106
*Vorticella*, 32, 35

wax, candle, 24, 61
  sealing, 15
wick, trimming of, 72

xylene, 60, 194ff

*Zea*, 141
zoom attachment, 135
zoophytes, 67
*Zygnema*, 32